Thierry Meyer, Genserik Reniers, Valerio Cozzani

Risk Management and Education

Also of interest

Security Risk Assessment
In the Chemical and Process Industry
Reniers, Khakzad, Van Gelder (Eds.), 2018
ISBN 978-3-11-050052-3, e-ISBN 978-3-11-049908-7

Engineering Risk Management.
Meyer, Reniers; 2016
ISBN 978-3-11-041803-3, e-ISBN 978-3-11-041804-0

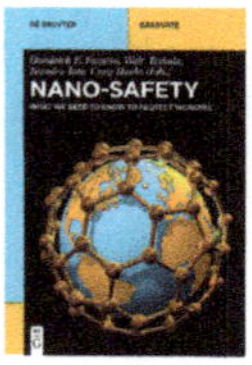

Nano-Safety
What We Need to Know to Protect Workers
Fazarro, Trybula, Tate, Hanks (Eds.) 2017
ISBN 978-3-11-037375-2, e-ISBN 978-3-11-037376-9

Research Laboratory Safety
Kuespert, 2016
ISBN 978-3-11-044439-1, e-ISBN 978-3-11-044443-8

Thierry Meyer, Genserik Reniers,
Valerio Cozzani

Risk Management and Education

DE GRUYTER

Authors
MER Thierry Meyer
Ecole Polytechnique
Fédérale de Lausanne
EPFL-SB-ISIC-GSCP
Station 6
1015 Lausanne
Suisse
thierry.meyer@epfl.ch

Prof. Valerio Cozzani
University of Bologna
LISES – DICAM
Via Terracini 28
40131 Bologna
Italy
valerio.cozzani@unibo.it

Prof. Genserik Reniers
Delft University of Technology
Safety and Security Science
Jaffalaan 5
2628 BX Delft
The Netherlands
g.l.l.m.e.reniers@tudelft.nl

ISBN 978-3-11-034456-1
e-ISBN (E-BOOK) 978-3-11-034457-8
e-ISBN (EPUB) 978-3-11-038377-5

Library of Congress Control Number: 2019931404

Bibliographic information published by the Deutsche Nationalbibliothek
The Deutsche Nationalbibliothek lists this publication in the Deutsche Nationalbibliografie; detailed bibliographic data are available on the Internet at http://dnb.dnb.de.

Typesetting: Integra Software Services Pvt. Ltd.
Printing and binding: CPI books GmbH, Leck
Cover image: hansslegers / iStock / Getty Images Plus

www.degruyter.com

Preface

Technological innovation is nowadays called to fulfill societal demand for new and high value-added products coping with the requirements of a sustainable development. The features of the markets and the urgency of a problem such as global warming, health, and well-being require to reduce the time to market new industrial processes and products. Thus, engineers are called more and more to assess and manage the associated risk under growing time pressures. Negative risk turning into an accident is a possibility that, if it occurs, adversely affects the ability of an undertaking to achieve its outcome objectives. Engineering risk management is the process of identifying risk, assessing risk, and taking steps to reduce it to an acceptable level. Engineering risk management can be defined as the early and continuous identification, assessment, and resolution of nonfinancial risks such as the most effective and efficient decisions can be taken to manage these risks.

A substantial body of knowledge has been developed around risk management, and some books dealing with engineering risk management have also been published, particularly by some of the authors of this book.

However, if the interest of considering engineering risk management is no more to be demonstrated, the way risk and safety concepts should be learned, taught, and acquired by the learners and engineers still remains of significant interest. Indeed, while the world has changed from an individual customer point of view, the way students (and also professionals in continuous training) learn has also moved a lot.

On the one hand, while the engineer's job evolves, the number of technical and nontechnical skills and knowledge to be covered by the learners during their training, and to master before graduation, has widely increased. In these conditions, with constant training duration, the teachers may have to spend less time on a given subject, and thus they need to be more efficient and make good choices to clearly deliver and illustrate the subject they want to cover.

On the other hand, new students learn differently. The Y generation, or digital natives, is characterized, among others, by its

- Connectivity and its ease with technological tools: They think "on the net" and know/feel that the more they are connected, the stronger they are. Social networks are also a part of everyday life.
- Impatience: They are always connected to the digital world; their access to information is instantaneous.
- Inventiveness: The improvement is no longer continuous, and intercultural settings and interdisciplinary solutions promote innovation.
- Different relationships with authority, including teachers: Authority is no longer related to a status, but must be demonstrated through competence and behavior.

https://doi.org/10.1515/9783110344578-201

In such conditions, the classical forms of teaching are no longer suitable, and new ways of teaching, based on the use of numerical devices, interactivity, active learning and teaching, group work, role playing, and so on have emerged and are successfully used by many teachers around the world. The main characteristic of these methods is to provide so-called active or interactive learning, where the learners are called to play an active role in the learning process.

It is thus a real pleasure to welcome and introduce such a book dealing with engineering risk management and its education. Thierry Meyer is professor of Chemical Engineering and Safety and Risk Management at the Federal Institute of Technology in Lausanne (EPFL), Switzerland, and Swiss academic delegate in the Working Party Education of the European Federation of Chemical Engineering. Genserik Reniers, chemical engineer by education, is also professor of Safety and Security Engineering and Management at both the KU Leuven. Valerio Cozzani is professor of Chemical, Engineering and Director of Academic Programs for the Department of Chemical, Environmental and Civil Engineering at the University of Bologna in Italy, also specialized in Process and Environmental Safety, and particularly involved in didactic interventions on Process Safety Engineering both in academia and in professional training.

By publishing this book, they now allow, being part of the scientific community, for the teachers and learners of different universities and levels, to access all their skills and knowledge on the teaching of engineering risk management.

An introduction to risk and safety management and their relation with chemical engineering is given in the first chapter. The need for engineering risk management education, from the regular education as well as a continuous training point of view, and the importance of education for prevention are then highlighted in the second chapter. The third chapter details the different education profiles for risk management, for both a classic risk manager profile and a continuing education profile, and it also covers the different forms of safety education. The different learning characteristics and outcomes are then fully described in the fourth chapter: what, when, to whom, how, should Engineering Risk Management be taught, and how should it be tested? This chapter also gives an overview of actual researches about risk management education and leaning outcomes, recent findings on safety education, perspectives about engineering risk management and different industrial sectors needs and expectations. Finally, examples of safety or risk management education programs, courses, and contents, for different universities and within different countries, are given in the fifth chapter, including illustrative courses, exams, and future thoughts. The book ends with conclusions and perspectives on risk management education, new technologies for education and networking, as well as competences and resources for education.

This complete and well-documented book takes advantage from all the academic, industrial, and international experiences of its authors. It will be useful to students, learners, teachers, and engineers who have to deal with engineering risk management and its education. I warmly thank Thierry Meyer, Genserik Reniers, and Valerio Cozzani for their important and original contributions to the subject.

Eric Schaer
Chairman or the Working Party Education (EFCE)

Contents

Preface — V

1 Introduction to risk management (RM) — 1
1.1 The RM process — **3**
1.2 RM and resources/financial aspects — **5**
1.3 RM and safety culture — **6**
1.4 The role of education — **9**
1.5 Safety management education — **11**
1.6 Conclusion — **12**
References — **13**

2 The need for engineering risk management (ERM) education — 15
2.1 Is ERM only a matter of professionals and professional training? — **15**
2.1.1 The third safety revolution: CHESS — **15**
2.1.2 Teaching engineering risk management (ERM) — **17**
2.1.3 Digitalization and education — **18**
2.2 Could we make an optimal prevention without education? — **18**
2.2.1 Educational approaches — **18**
2.2.2 Person- and behavior-based safety — **19**
2.2.3 Improved KPABC model — **20**
2.2.4 Safety as a separate education subject – and not — **21**
2.3 Standards and education — **22**
2.3.1 Standards in the education of engineering risk managers (ERManagers) — **22**
2.3.2 Competence required by the application of standards in ERM — **24**
2.4 Conclusion — **26**
References — **26**

3 Education profiles for ERM — 29
3.1 Different education profiles for ERManagers — **29**
3.2 Elements for education of the “classic risk manager profile” — **32**
3.2.1 The process industries as an example — **33**
3.2.2 The classic risk manager profile — **37**
3.3 Elements of education of a continuing education profile — **38**
3.4 Different forms of safety education — **41**
3.5 Conclusion — **49**
References — **49**

4 **Learning characteristics and outcomes — 51**
4.1 Safety teaching and training — **51**
4.2 Process safety education (as an example) — **53**
4.2.1 Introduction — **53**
4.2.2 Process safety accreditation — **55**
4.2.3 Process safety curricula — **56**
4.2.4 Teaching strategies for process safety — **56**
4.2.5 Process safety research fields — **57**
4.2.6 Process safety collaboration strategies — **58**
4.2.7 Process safety competence — **60**
4.2.8 Process safety training — **61**
4.2.9 Process safety education: a discussion — **62**
4.3 Conclusion — **63**
References — **64**

5 **Example of safety or ERM education programs, courses and contents — 69**
5.1 University programs — **69**
5.1.1 Skills required in the educational background of ERManagers — **69**
5.1.2 Competence providers — **69**
5.1.3 General framework of university programs for ERM education — **71**
5.2 Different examples from different countries — **73**
5.2.1 Industrial risk management — **74**
5.2.2 Reliability and safety management — **77**
5.2.3 Process safety management — **78**
5.2.4 Disaster risk management — **80**
5.2.5 Environmental risk management — **81**
5.3 Illustrative courses — **84**
5.4 Conclusions — **84**
References — **88**

6 **Conclusions and perspectives — 89**
6.1 Perspectives on risk management education — **89**
6.2 New technologies and new delivery formats for education and networking — **90**
6.3 Competences and resources for education — **95**
References — **99**

Acronyms — 101

Index — 103

1 Introduction to risk management (RM)

Trying to define what is a risk is generally compartmentalized based upon whether the risk is in the context of business continuity, project management, security, engineering, industrial processes, financial portfolios, actuarial assessments, or health and safety. The potential list is finite, but is certainly overwhelming. Risks could be described as disruptions resulting from the unpredictability of the future caused by accidental derogation possibilities of planned targets. Therefore, talking about risks also means the dispersion around an expected value.

The authors want to emphasize that this book is concerned with operational risks and not with financial ones. The latter refer to a company's ability to manage its debt and financial leverage. It can also be seen as the probability of loss (or profit) inherent in financing methods that may impair the ability to provide adequate return, or like any risk that comes from giving money to another person or entity. For example, if one lends money, one carries the financial risk that the borrower will not repay it. A venture capital firm carries the financial risk that its investments will never become profitable. Likewise, an investor who purchases an asset carries the financial risk that he/she will be unable to resell it.

Operational risk is the risk not inherent in financial, systematic, or market-wide risk. It is the prospect of loss resulting from inadequate or failed procedures, systems, or policies, such as

- employee errors;
- systems failures;
- fraud or other criminal activity;
- any event that disrupts business processes;
- reputation problems.

Most organizations accept that their people and processes will inherently incur errors and contribute to ineffective operations. In evaluating operational risk, practical remedial steps should be emphasized in order to eliminate exposures and ensure successful responses.

Risk is part of all our lives. As an individual, we daily take risks even if we are not aware of it. If you went to work this morning, you took a risk. If you rode your bicycle, used public transportation, walked, or drove a car, you took a risk. If you put your money in a bank, in stocks, or under a mattress, you took other types of risk. If you bought a lottery ticket, you were involving an element of chance – something intimately connected with risk. The latter example implies that in specific domains, in particular, when dealing with financial or technological opportunities,

https://doi.org/10.1515/9783110344578-001

risk not always has a negative accent, but may be interpreted as an uncertainty affecting the achievement of a given objective.

This also explains why, at another scale, as a society, we need to take risks to grow and develop. From energy to infrastructure, supply chains to transportation security, hospitals to housing, effectively managed risks help societies achieve growth and development. In our fast-paced world, the risks we have to manage evolve quickly, nowadays, even quicker. We need to make sure we manage risks so that we minimize their threats and maximize their potential.

In their book *Engineering Risk Management*, Meyer and Reniers [1] depicted risk management (RM) with an engineering perspective. RM involves understanding, analyzing, and addressing risk to make sure organizations achieve their objectives. So it must be proportionate to the complexity and type of organization involved. RM is an integrated and joined up approach to managing risk across an organization and its interrelations with the extended networks. It must be analyzed using a systemic concept. Because risk is inherent in everything we do, it could address diverse topics such as health and safety, engineering, planning, finance, insurance, banking, business continuity, financial services, politics, and many more.

Being risk sensitive is not the same as being paranoid. One can realize that there are risks associated with everything. There is a need to take a deliberate and methodical approach to dealing with risk, while at the same time being realistic. The central idea of risk is that there is uncertainty involved. If compared with life [2], the only certainty in life is death, and the uncertainty lies in when and how death will occur. People strive to delay the final outcome of life and try to improve the quality of life in the interim. Threats to these interim objectives involve risks, some natural, some man-made, some completely beyond our control, but most of them are controllable and manageable.

Risk appetite is the amount of risk, on a broad level, an organization is willing to accept in pursuit of value. Each organization pursues various objectives to add value, and should broadly understand the risk it is willing to undertake in doing so [3]. No organization can achieve its objectives without taking risks, but the level and amount of risks an organization has to take cannot be clearly specified. The biggest challenge is to recognize the risks and to manage them continuously. It should be remarked that risk is not static, but has an inherently dynamic feature, since it is often affected by a huge number of external factors likely to evolve in time.

RM is not limited to catastrophic failures of assets or processes. To be effective, RM must acknowledge that risks take many forms and that all must be clearly understood and effectively managed. Finally, RM is multidimensional and requires the direct support of most business and plant functions, as well as the entire workforce of an organization.

1.1 The RM process

What does RM mean? Is it just identification, assessment, and planning and controlling social, economic, and/or physical threats to an organization? Is the concept only about transferring the risk or reduce its negative effects? With no doubt, the answers for the above questions is "no."

The process of RM is not only restricted to controlling the threats or reducing their negative effects. It is a much deeper concept that also involves risk avoiding as well as risk taking. Every work involves some kinds of risk. Sometimes you avoid, sometimes you control the phenomenon, and sometimes you simply let it come.

We will focus on the engineering approach for the RM process. The International Organization for Standardization (ISO) in ISO 31000:2018 [4] identifies the following principles of RM:

- Organizations of all types and sizes face external and internal factors and influences that make it uncertain whether they will achieve their objectives.
- Managing risk is iterative and assists organizations in setting strategy, achieving objectives, and making informed decisions.
- Managing risk is part of governance and leadership, and is fundamental to how the organization is managed at all levels. It contributes to the improvement of management systems.
- Managing risk is part of all activities associated with an organization and includes interaction with stakeholders.
- Managing risk considers the external and internal context of the organization, including human behavior and cultural factors.

The process could be divided into seven main steps as follows:
Step 1 – *RM objectives*: In order to effectively identify risk, a company should first at least define strategic, operational, reporting, and compliance objectives. The purpose of establishing the scope, context, and criteria is to customize the RM process, enabling effective risk assessment and appropriate risk treatment. Scope, context, and criteria involve defining the scope of the process, and understanding the external and internal context.

Step 2 – *Risk identification*: Recognize and describe risks that might affect the project or its outcomes. What is the extent of risk faced? What are the available options? How large, and how immediate are the outcomes resulting from the impact of risk? Can the risk be controlled, reversed, or avoided? How do individuals and groups conceptualize risk? What aspects of the problem seem most relevant?

Step 3 – *Risk analysis*: Once risks are identified, they should be quantified or at least estimated in terms of likelihood and severity (damages). Develop an understanding of the nature of the risk and of its potential to affect project goals and objectives.

Step 4 – *Risk evaluation*: Evaluate and/or rank the risk by determining the risk magnitude, which is the combination of likelihood and consequence (being the simplest model). You make decisions about whether the risk is acceptable or whether it is serious enough to warrant treatment. Companies must first prioritize risks to identify and limit them and then further prioritize and address the rest of the risks based on the needs of the organization.

Step 5 – *Risk treatment*: Selecting the most appropriate risk treatment option(s) involves balancing the potential benefits derived in relation to the achievement of the objectives against costs, effort, or disadvantages of implementation. During this step, assess the highest ranked risks and set out a plan to treat or modify these risks to achieve acceptable risk levels. How can the probability of the negative risks be minimized, as well as the opportunities that go hand in hand with those risks be enhanced? Create risk mitigation strategies, preventive plans, and contingency plans in this step.

Step 6 – *Monitor and Review the risk*: The purpose of monitoring and review is to assure and improve the quality and effectiveness of process design, implementation, and outcomes. Ongoing monitoring and periodic review of the RM process and its outcomes should be a planned part of the RM process, with responsibilities clearly defined.

Monitoring and review should take place in all stages of the process. Monitoring and review includes planning, gathering and analyzing information, recording results, and providing feedback. The results of monitoring and review should be incorporated throughout the organization's performance management, measurement, and reporting activities.

Step 7– *Recording and reporting*: The RM process and its outcomes should be documented and reported through appropriate mechanisms.

To create effective RM, appropriate controls or countermeasures to measure each risk must be selected. Risk mitigation needs to be approved by the appropriate level of management. For example, a risk concerning the image of the organization should have top management decisions behind it. If it is a technical issue, for example, a fire safety device, the risk could be mitigated by acquiring the adequate equipment depending on the fire department or engineering facilities.

A further element is that risk should not be intended as a static, but as a dynamic concept. Changes in risk may derive from updates in the system or in the procedures adopted, in the facilities, and in the training of workers. Moreover, new knowledge or new awareness may become available, for example, due to scientific research or due to lessons learnt from past events, leading to the identification of new risks that need to be controlled. Thus, the above procedure needs to be repeated in time, periodically updating the risks to be controlled and the RM system.

RM should propose applicable and effective safety and security controls for managing the risks. It should contain a schedule for control implementation and responsible persons for those actions.

Therefore, the last rule or step would be to use the Japanese Kaizen approach [5]: measure the effects of your RM efforts and continuously implement improvements to make it even better.

1.2 RM and resources/financial aspects

All risks can never be fully avoided or mitigated simply because of financial constraints, as well as of limited resources and practical problems. Therefore, all organizations have to accept some level of residual risks, but it is imperative that all risks are isolated, clearly defined, and managed within financial and practical constraints.

RM also faces difficulties in allocating resources (money, manpower, facilities, etc.). This is the idea of opportunity cost. Resources spent on RM could have been spent on "more profitable" (at least, at first sight) activities. Again, RM should tend to minimize spending and minimize the negative effects of risks. Optimizing prevention investments and making investment decisions in a cost-efficient way is therefore essential for corporations. For example, Reniers and Sörensen [6] suggest a user-friendly knapsack-based approach to take cost-efficient prevention decisions. Prevention costs are weighed against hypothetical benefits following the preventive measures taken, and the most cost-efficient preventive measures are determined following the knapsack algorithm, given a certain prevention budget available.

Risk-based resource allocation is more important now than ever, for several reasons. First, as societies have matured, we have become ever more insistent on living in safe and predictable conditions. As such, the number of laws and regulations has increased, expanding the scope of necessary enforcement. Second, we demand ever more efficiency from companies and government. Thus, regulatory and enforcement agencies face pressure to fulfill their missions with ever tighter budgets.

Some keys to success could be expressed as follows:

- Implement thoughtful controls: in principle, the best-designed controls are capable of preventing future violation, and so they are often quite efficient.
- When adverse events are difficult to forestall, as in the case of natural hazards such as earthquakes do, focus on reducing vulnerabilities or on mitigating the potential damage.
- Allocate detection efforts based on risk. One should focus resources wherever the risks are the greatest. It is a powerful driver of efficiency and effectiveness, and often only few people focus their efforts in this way.

- When potential threats are sophisticated, focus more deeply on fewer targets. Concentrating the efforts on a smaller target is worth while when multiple threats are present.
- When people who are inspected could remediate violations quickly, use random inspections. Here, one has to find a balance between efficiency and effectiveness. It is quite always more efficient for audits to be planned in advance, so that the inspected party can have the requisite information ready and have cleared time to interact with the auditors. On the other hand, if they know that an auditor is coming, they will generally make every effort to conceal or remediate bad behavior, making the inspection less effective.
- Inspect both roughly and deeply. One should measure a given set of variables consistently over time to get a holistic picture of activities and to enable it to draw inferences about how behavior may be changing.
- Educate inspected people who have good intentions. When it is reasonable to judge that violations were unintentional, there is an opportunity for auditors to improve compliance through *education*, by making instructions clearer or by notifying the parties of their noncompliant behavior [7].

1.3 RM and safety culture

The industrial revolution in the nineteenth century radically increased the quality of life for society.

However, the fast development of new manufacturing processes was closely intertwined with the creation of new dangers, and only major catastrophes triggered the motivation to understand and manage them [8]. Each catastrophe either revealed safety deficiencies that had to be addressed or, in some instances, indicated that lessons were not learned and further stressed the importance of dealing with them. As new approaches to tackle deficiencies in safety were created and perfected, the rates of accidents reduced over time but eventually plateaued [9–10]. The drive to increase safety has, however, never dwindled, because the increase of public intolerance with respect to industrial accidents implies that accident rates will never reach satisfactory levels.

The best example to illustrate the concept of safety culture and how it emerged including the relationship it has with RM is to take the instance of the perspective of the nuclear power plant (NPP) industry. N. Gerweck [11] in her internal report depicted this perspective using three major accidents, from Three Mile Island (TMI) over Chernobyl to Fukushima. Some extracts are reprinted with permission here.

Initially, the approach to enhance safety was to make technological improvements. For example, in the nineteenth century, boilers on trains exploded regularly until safety valves were introduced [12]. As accidents with technical deficiencies as a root cause diminished drastically, accidents with human-related root causes became apparent and revealed the importance of the human role. This point can be illustrated with the TMI NPP accident in 1978, which resulted in about half of the fuel core melting. While there were no casualties arising from the accident, its impact was tremendous on the US NPP industry: it took the US Nuclear Regulatory Commission (USNRC) over 30 years to grant a new license to build a new NPP [13].

On the 28th of March 1978, the TMI number 2 reactor (TMI-2) shut down automatically due to a relatively minor malfunction. However, the operators, unable to diagnose the cause of the problem, overrode the automatic safety systems and performed manual interventions that aggravated the situation. Had the operators not intervened and let the automatic safety systems function on their own, the major catastrophe would have remained a minor incident [14].

Whilst this accident had experts consider the human role on an individual level, for example, recognizing the importance of elements such as training and proper human–machine interfaces, it was the disaster that occurred at the Chernobyl NPP in 1986 that formally introduced the concept that individuals alone cannot be blamed, and that the context in which they operate must be considered.

On the day prior to the Chernobyl NPP disaster, the operators were instructed to perform a test that required the nuclear reactor to be set up in a dangerous configuration. However, they were not alerted of the safety implications and of the potential dangers. The operators were under increasing pressure to finalize the test because it had already been performed previously with inconclusive results. Furthermore, as the operators were falling behind schedule, instead of being completed in one shift as originally intended, the test ended up being spread over three, the biggest part of the testing falling upon the night shift, which was run by unprepared operators who had not expected to carry out the unusual procedure. When the test did not go as planned, rather than cancelling it, the operators modified the protocol, which exacerbated the already precarious conditions of the plant. As a result, the reactor exploded releasing large quantities of radioactive particles that spread across Europe and launched an international crisis.

Although it was the operators who triggered the accident by violating the testing protocols that had been provided to them, it was recognized that they alone could not be blamed: lack of accountability, pressure from above to perform unsafe experiments, the impossibility to question or refuse to carry out orders, and a blatant disregard for safety were the contributing factors that led to the disaster. The International Nuclear Safety Advisory Group (INSAG) coined the term "safety culture" to be able to explain the root cause of the Chernobyl NPP accident. Although it did not provide a definition at the time, the USNRC has since defined safety culture as "the core values and behaviors resulting from a collective commitment by leaders and individuals to emphasize safety over competing goals to ensure protection of people and the environment."

The report of the INSAG stated that:

> The (Chernobyl) accident can be said to have flowed from deficient safety culture, not only at the Chernobyl plant, but throughout the Soviet design, operating and regulatory organizations for nuclear power that existed at the time. Safety culture (...) requires total dedication, which at NPPs is primarily generated by the attitudes of managers of organizations involved in their development and operation. An assessment of the Chernobyl accident demonstrates that a deficit in safety culture was inherent not only at the stage of operation, but also to important activities at other stages in the lifetime of NPPs (including design, engineering, construction, manufacture and regulation). [15]

In 2011, another major NPP accident occurred, this time in Japan. Following the largest earthquake recorded in the country (M.9.0) since the beginning of the recording (year 684 [16]) and the subsequent tsunami that ensued, the Fukushima Daiichi NPP experienced a catastrophic meltdown. However, as stated by the chairman of the Fukushima nuclear accident independent investigation commission (NAIIC), "(The accident) cannot be regarded as a natural disaster. It was a profoundly manmade disaster – that could and should have been foreseen and prevented. (...) Japan's nuclear industry managed to avoid absorbing the critical lessons learned from Three Mile Island and Chernobyl" [17].

In stark contrast to the terrible fate of Fukushima Daiichi, the Onagawa NPP managed to shut down safely and was remarkably undamaged [18], even though it was positioned 60 km closer to the epicenter of the earthquake and the tsunami was higher by over a meter than when it struck Fukushima Daiichi. A notable testimony of the two divergent outcomes of the NPPs is the fact that people living near Fukushima Daiichi were ordered to evacuate, while residents from the town of Onagawa fled to the Onagawa NPP for shelter [19].

The main difference that resulted in the two outcomes was the safety culture of the companies operating the Fukushima Daiichi and Onagawa NPP, that is, Tokyo Electric Power Company (TEPCO) and Tohoku Electric Power Company, respectively. The diverging safety culture was already apparent from when the NPPs were in construction: instead of building the Fukushima Daiichi NPP 35 m above sea level, TEPCO elected to remove 25 m of the natural seawall to more easily transport equipment to the site and pump seawater to the reactors, in effect building the power plant at the much lower elevation of 10 m. On the other hand, the vice president of Tohoku Electric Power Company at the time of the construction of the Onogawa NPP, Yanosuke Hirai, strongly pushed to have a breakwater of 14.8 m built, a higher and more expensive wall than initially proposed. In the words of the NAIIC, since the construction of Fukushima Daiichi, "rather than considering the known facts and quickly implementing counter measures, TEPCO resorted to delaying tactics, such as presenting alternative scientific studies and lobbying" [17]. In contrast, Tohoku Electric Power Company was constantly looking to increase their knowledge and test the resiliency of the plant to guarantee its safety.

Since the Chernobyl accident, the concept of safety culture has been used and applied outside of the nuclear power plant industry, such as by the chemical industry, healthcare [20], and more recently, academia [21].

Research on how to instill a proper safety culture and how to measure its effectiveness is ongoing today. However, all the indicators tend to reveal that education is a key factor. Bringing the impact of its poor safety culture beyond the accidents that occur in research and teaching laboratories, academia has been held responsible for a part of the accidents that still occur in industries [22, 23].

1.4 The role of education

By the term education, we will encompass the professional education as well as the public education also being an important player in the RM process. Although public education regarding accidents and disasters has historical roots, education has not always followed or even preceded the hazard awareness. One of the questions to be answered is how we could be aware of a hazard or a hazardous situation if we have not been educated or trained about it? Education is a continuing process for improving the level of knowledge including the process of learning and acquiring information. It is widely believed that constant exposure to new ideas and skills makes people better workers, thinkers, and societal contributors [7]. As the traditional "goodwill" of people and staff is not sufficient to ensure a safety level, it is necessary to implement a safety management system including education.

Awareness of hazards and vulnerabilities will not, by itself, necessarily produce risk reduction behavior. Risk reduction education/capability is a necessary and crucial component of a safety culture. Risk reduction education seeks to educate and engage people in undertaking effective preventive and mitigation measures that can reduce exposure and vulnerability hazards. Knowledge of efficient and effective risk reduction strategies and measures, including possible financial risk-sharing and risk-transfer arrangements, is an important element of risk reduction education for individuals, businesses, civil society, and governments. This risk education also has the capacity to enhance awareness and use – at the individual, industrial, and governmental level – financial protection tools that can facilitate physical and economic recovery from hazardous events [24].

These tools include, for large damages, private insurance and reinsurance markets, dedicated catastrophe insurance and reinsurance programs and reinsurance arrangements, insurance pools, and alternative financial market solutions (e.g., catastrophe bonds).

The key to reducing loss of life, personal injuries, and damage from accidents or disasters (industrial, economic, or natural) is widespread public awareness and education. People must be made aware of what hazards and associated threats they are likely to face in their own workplace, environment, or communities. They should know in advance what specific preparations to make before an event (prevention), mitigation measures to implement (protection), what to do during an unexpected event or a crisis (reaction), and what actions to take in its aftermath (restore).

The OECD defines in its policy handbook on natural hazard awareness and disaster risk reduction education [25] nine policy guidance principles:

1. Natural hazard awareness and disaster risk reduction education are a *foundation and prerequisite* for effective catastrophic RM strategies at country and regional levels.
2. Risk awareness and risk reduction education priorities should be *risk based*, tailored to the hazards of the region and the particular vulnerabilities and capacities of those exposed to risk.
3. *Hazard mapping* and *risk assessments* provide the basis for elaborating risk awareness and risk reduction education strategies and for informing their content.
4. Natural hazard awareness and disaster risk education efforts should aim at encouraging *voluntary risk reduction activities*, thereby building support for a shift in disaster management and enhancing public acceptance of any necessary compulsory measures.
5. *Disaster risk transfer and financing mechanisms* can provide a mechanism for enhancing risk awareness and risk reduction education.
6. *Appropriate risk communication techniques* should be adopted to reach the targeted audiences and induce the desired changes in behavior and perception. Any risk reduction strategies contained in messages should be *specific and realistic for local conditions*.
7. Clear and consistent messages to all interested parties (including all levels of government) concerning the *allocation of expected disaster costs and disaster prevention responsibilities* can promote a shared understanding of roles and responsibilities and stimulate individual and collective actions to reduce vulnerability and exposure to the risk of physical and financial losses from natural hazards. Insurance supervisors should ensure that there is transparency and clarity in insurance contracts and industry communications concerning the scope of coverage, so as to avoid any confusion regarding the allocation of disaster risks between insurance companies and policyholders.
8. Promoting a culture of safety requires a *long-term and sustained strategy* by governments and must be based on a strong commitment of all institutional actors. To ensure continuity and sustainability, risk awareness and risk reduction education efforts should be well integrated into a cross-sectoral, national strategy for disaster risk reduction.
9. *Continuous monitoring* and *periodic evaluations* of awareness and education efforts should be conducted to assure accountability and transparency, and increase public confidence in the outcomes.

While increased awareness of industrial, emerging, or natural perils and disaster risk education may induce people to change their behavioral patterns and adopt riskwise conducts, it may also contribute to change their perceptions, thereby reinforcing support for, and legitimacy of, people or public initiatives such as the investment in more resilient infrastructures, safer behaviors, or compulsory measures. Moreover, it is only through education that it is possible to build public acceptance for stricter regulations and insurance programs that may be instituted.

1.5 Safety management education

Another aspect of the global RM concept is to include safety management, as previously mentioned. Safety education is one of the most important aspects of safety management. One of its aims is to help in defining the curriculum of postsecondary institutions to ensure that future business managers and leaders create healthier and safer workplaces.

A key point of the efforts to improve safety practices is enhancing the culture of safety [26]. However, as previously mentioned, safety culture cannot be sustained without education. Continuous and spiral safety education [27] reinforces the value and need for safety, and demonstrates to students the high priority that employees and managers should place on safety.

Teaching safety over a long period of time builds the knowledge base and the safety skills the students will need as they pursue further education and move into the workforce. Continuous safety education may, for instance, be accomplished by including a relevant safety topic during certain courses (such as laboratory sessions). Teaching safety in a variety of topics can be challenging at times, since there are many types of hazards, many types of workplaces, for example, laboratories, and many types of work. Moreover, it becomes challenging to train or teach safety when accidents are quite rare [7]. Therefore, building a strong base of knowledge and skills in safety throughout an undergraduate study requires teaching numerous safety topics.

It should be noted that a concern about a single course is that there is an accompanying sense that education on this topic has finished upon completion of the course – learning safety should be a continuous process.

Within a strong safety culture, safety knowledge should be built year after year throughout the undergraduate curriculum and continue into graduate (and even postgraduate) studies, so that upon graduation, graduates have superior safety skills and strong safety ethics. It creates a solid foundation upon which a successful health and safety program can be built in a working environment. As part of that culture, all levels of an education-related organization (i.e., administrative personnel, scientists, laboratory technicians, and students) should understand the importance of minimizing the risk in a working environment and should work together toward this end. For example, lab personnel should consider the health, physical, and environmental hazards of the products/processes that will be used when planning a new experiment and perform their work in a safe manner. However, the ability to accurately identify and assess hazards at the workplace is not a skill that comes naturally, and it must be taught (through classical and continuing education) and encouraged through training and ongoing organizational support. A successful safety program requires a daily commitment from everyone in any organization, and setting a good example is the best method of demonstrating commitment.

Training in safety practices is an essential element in developing and sustaining this safety culture. This is particularly important in an environment where staff is relatively young and unexperienced. They do not know how to assess the risks they face, and/or how to assess the changes in risks when conditions are modified [28].

Within a setting of higher education, starting from safety in laboratory practices, the concept of safety may be widened in the education of future professionals in academic courses. Ethics and social responsibility may be as well an important part of such education. Besides their own safety, students need to be trained in recognizing the social responsibility they will have when moving to positions in companies. Applying safety criteria to the operation of laboratories or of industrial processes, to the design, and to the operation of industrial systems and components needs to be assumed as a responsibility on the workplace and toward society. Safety education and examples of safety commitment during academic training have a crucial role in this perspective.

Geller [29] suggests that a safety culture is composed of three domains, including environmental, personal, and behavioral factors, and his model focuses on changing the behavior and the mind-set of the individual. Thus, to develop a strong safety culture all stakeholders need to understand why safety rules are valuable. If the first two domains could be supported by education, this is more difficult for the behavioral component. Ultimately, an ideal behavioral component would lead to a safety culture where proactive and immediate action of every aspect of safety that requires attention becomes second nature to every stakeholder [30]. Under these conditions, compliance, and even behaving beyond compliance, becomes spontaneous, as it becomes a byproduct of a good safety culture.

1.6 Conclusion

Nowadays, ERM is a general and well-accepted concept, but its implementation is sometimes lacking. Going back to the root cause, we could at least identify one key factor, being education of all the involved stakeholders, to be essential.

Over time, public's expectation regarding safety has been increasing, such that accidents with smaller consequences have also started to generate indignation.

Regulators have also straightened the rules, and the constraints for the companies and hence for the employees have become stricter.

Perhaps the public's increasing expectations for the safety of students and workers, and the industry's increasing expectations regarding prospective employees, might provide enough pressure to help the proponents to be able to change the education system and its mentality, including academia.

Risk awareness and education efforts should place emphasis on concrete risk reduction tools and strategies that can be adopted. Moreover, to be fully effective and efficient, these efforts should take place at, and be targeted to, every layer of society: at the individual, business, civil society, and governmental levels.

References

[1] Meyer, T., Reniers, G. (2016) *Engineering Risk Management*, Berlin, Walter de Gruyter GmbH.

[2] Rowe, W.D. (1977) *An Anatomy of Risk*. New York: John Wiley & Sons Inc.

[3] Rittenberg, D. L., Martens, F. (2012) *Enterprise Risk Management – Understanding and Communicating Risk Appetite*. Durham: The Committee of Sponsoring Organizations of the Treadway Commission (COSO).

[4] International Organization for Standardization, ISO 31000:2018 (2018) Risk management – Guidelines, ISO norms.

[5] Imai, M. (1986) *Kaizen: The Key to Japan's Competitive Success*. New York: Random House.

[6] Reniers, G.L., Sörensen, K. (2013) An approach for optimal allocation of safety resources: using the knapsack problem to take aggregated cost-efficient preventive measures, *Risk Analysis*, 33(11),2056–67.

[7] Meyer, Th. (2017) Towards the implementation of a safety education program in a teaching and research institution, *Education for Chemical Engineers*, 18, 2–10.

[8] Mannan, M. S., Reyes-Valdes, O., Jain, P., Tamim, N., Ahammad, M. (2016) The Evolution of Process Safety: Current Status and Future Direction, *Annual Review of Chemical and Biomolecular Engineering*, 7, 135–162.

[9] Daniellou, F., Simard, M., Boissière, I. (2011) *Human and Organizational Factors of Safety: State of the Art*, FonCSI.

[10] Pasman, H., Suter, G. (2005) EFCE Working Party on Loss Prevention and Safety Promotion, *Process Industries Process Safety and Environmental Protection*, 83, 18–21.

[11] Gerweck N. (2017) The evolution of safety in industry and academia, EPFL internal report.

[12] Morse, G. (2014) Railway Accidents, Bloomsbury Publishing PLC, London.

[13] US Energy Information Administration (2012) Nuclear Regulatory Commission approves construction of first nuclear units in 30 years https://www.eia.gov/todayinenergy/detail.php?id=5250.

[14] Erp, Jan B. van (2002) Safety culture and the accident at Three Mile Island (IAEA-CN--97). International Atomic Energy Agency (IAEA).

[15] International Atomic Energy Agency (1992) The Chernobyl Accident: Updating of INSAG-1 International Atomic Energy Agency.

[16] S. L. Soloviev and Ch. N. Go, (1974) Catalogue of tsunamis on the western shore of the Pacific Ocean, Nauka Publishing House, Moscow; translated in Canadian Translation of Fisheries and Aquatic Sciences No. 5077.

[17] Kurokawa, K., Ishibashi, K., Oshima, K., Sakiyama, H., Sakurai, M., et al. (2012) The National Diet of Japan Fukushima Nuclear Accident Independent Investigation Commission. NAIIC.

[18] International Atomic Energy Agency (2012) IAEA Mission to Onagawa Nuclear Power Station to Examine the Performance of Systems, Structures and Components Following the Great East Japanese Earthquake and Tsunami. IAEA Mission Report, Department of Nuclear Safety and Security, IAEA.

[19] The guardian (2011), Onagawa: Japanese tsunami town where nuclear plant is the safest place https://www.theguardian.com/world/2011/mar/30/onagawa-tsunami-refugees-nuclear-plant.

[20] Relihan, E., Glynn, S., Daly, D., Silke, B., Ryder, S. (2009) Measuring and benchmarking safety culture: application of the safety attitudes questionnaire to an acute medical admissions unit, *Irish Journal of Medical Science*, 178, 433–439.

[21] Hill Jr, R.H. (2012) Creating Safety Cultures in Academic Institutions, *Chemical and Engineering News*, 90, 24–46.

[22] Louvar, J. F., Hendershot, D. C. (2003) SACHE: 17 years of promoting teaching of safety to chemical engineering students. *Chemical Health and Safety*, 10, 8–10

[23] Hendershot, D. C., Smades, W. (2007) Safety culture begins in the classroom, *Process Safety Progress*, 26, 83–84.

[24] Green, R. et M. Petal (2009), Stocktaking Report and Policy Recommendations on Risk Awareness and Education on Natural Catastrophes, OECD Journal: General Papers, vol. 2008/3, https://doi.org/10.1787/gen_papers-v2008-art21-en.

[25] OECD (2010) Policy Handbook on Natural Hazard Awareness and Disaster Risk Reduction Education, OECD, France.

[26] Cox, S., Cox, T. (1991) The structure of employee attitudes to safety – a European example, *Work and Stress*, 5, 93–106.

[27] Robert h. Hill, David S. Crumrine, Laurence J. Doemeny, Steven A. Fleming, Kenneth P. Fivizzani, Douglas B. Hausner, Kirk P. Hunter, Edward M. Movitz, George S. Wilson, Marta U. Gmurczyk, (2012) Creating Safety Cultures in Academic Institutions, American Chemical Society, Washington.

[28] National Research Council (2014) *Safe Science: Promoting a Culture of Safety in Academic Chemical Research*. Washington, DC: The National Academies Press.

[29] Geller, E. S. (1996) *The Psychology of Safety: How to Improve Behaviors and Attitudes on the Job*. Chilton Book Company: Radnor, PA.

[30] Staehle, I.O, Chung, T.S., Stopin, A., Vadehra, G.S., Hsieh, S.I., Gibson, J.H., Garcia-Garibay, M.A. (2016) An Approach to Enhance the Safety Culture of an Academic Chemistry Research Laboratory by Addressing Behavioral Factors, *Journal of Chemical Education*, 93 (2), 217–222.

2 The need for engineering risk management (ERM) education

2.1 Is ERM only a matter of professionals and professional training?

Nowadays companies recognize the need to train and develop their risk management staff; corporate training programs will increasingly feature engineering risk management. These training programs will likely be a combination of internal and external resources, and include internal workshops, external conferences, and Internet-based training tools.

Despite the evolution of safety over the past century, which is based on publications by Swuste et al. [1–4ww] (see Figure 2.1), current approaches and contemporary thinking, mental models, technological approaches and solutions, safety implementations, and (evolutionary) ways to improve safety are regretfully not sufficient to revolutionize safety and making society and industrial activities much safer in a realistic and achievable way. Therefore, a further safety revolution is needed.

In fact, two safety revolutions already took place on a global scale (Figure 2.1): (i) the "safety first movement" (1900s until 1950s) represents the first safety revolution, and (ii) the "risk management and loss prevention" approaches (1960s until 2010s) denote the era of the second safety revolution. The third safety revolution that is needed today to further advance safety includes, among others, education on engineering risk management (ERM) on a much wider and more sustainable scale than is the case today. Figure 2.1 shows the three safety revolutions along with the underlying theories, models, concepts, and ideas per decade.

Revolutions start with radically new ideas. These new ideas are formed via mental models, the willingness to change things, and the understanding that changing things will lead to an improved situation, which in turn will result in increased profitability of industry and augmented benefit for society. Such should be the case with the third safety revolution.

2.1.1 The third safety revolution: CHESS

Reniers and Khakzad [5] believe that the third safety revolution can be represented by the acronym "CHESS." The authors explain that CHESS in fact summarizes five very important fields where revolutionary progress is needed:
- Cluster-thinking and intensified cooperation
- High transparency and efficient inspection
- Education, training, and learning

https://doi.org/10.1515/9783110344578-002

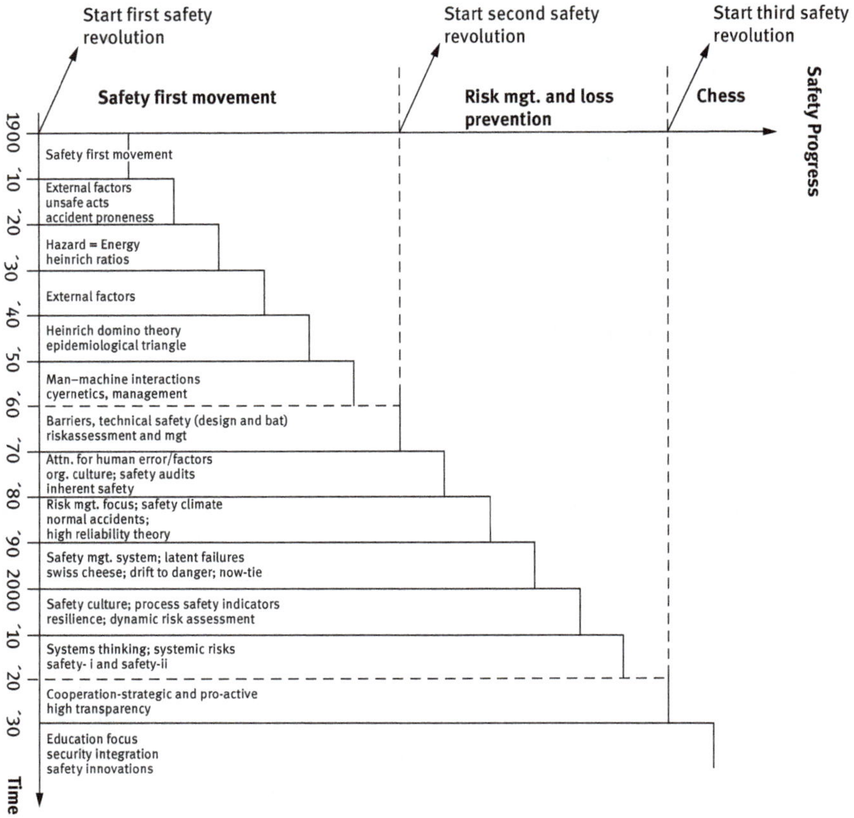

Figure 2.1: Safety progress and safety revolutions in the chemical industry (1900–2030 and future). Source: Reniers and Khakzad [5].

- Security integration
- Safety innovation and dynamic risk assessment

The authors further indicate that at first sight, these fields represent well-known recipes for improving safety in any industry, whereas they are nothing new. However, the combination of these domains will indeed lead to a third safety revolution if they will be addressed in radical innovative ways. The required innovation can be exemplified by a number of concrete ideas, which can only be realized if the current mentality of practitioners, academics, and authorities changes. For further information, we refer to the article by Reniers and Khakzad [5].

With respect to the third revolutionary field, dealing with safety education, learning, and training, it is argued that an individual not only needs to learn from near misses and incidents (reactive learning) but also from safety models, theories, and knowledge in general (proactive learning). The authors defend the viewpoint that there lays a task for society: there should be courses on "dealing with risk and uncertainty," or "operational safety," starting from primary school education. If people get familiar with safety from very early ages, they can learn much more in higher education and in professional life. Moreover, it can be expected that safety knowledge of people through regular education will be translated into daily life use and daily business use, to make better decisions and reduce losses, both on private and public working levels.

2.1.2 Teaching engineering risk management (ERM)

Nonetheless, it should be acknowledged that teaching safety and ERM is quite different from teaching an academic subject. Education with respect to ERM not only consists in merely providing information, but also in changing a mental attitude. This is always difficult, since students (and people in general) resent being preached to. In this regard, it is evident that it may be much easier to educate professionals on this topic than to educate teenagers or those in early twenties without any professional background.

In its most simple form, ERM teaching requires students, whether they are professionals or not, to be shown how to recognize dangers, and to be encouraged to make personal suggestions for the improvement of "avoiding losses" and the organizing of safety (including security). When every person would act as his/her own loss prevention manager, the highest safety and performance level would be achieved in a most efficient way. It is not possible to achieve this goal in an optimal way only by teaching and training professionals. ERM and operational safety should be a mental disposition built up from early schooldays, thus becoming part of decision-making of any person.

Moreover, if the basics of ERM and safety would already have been taught in primary school and in high school, it is possible for higher education to go much further and deeper into specific loss prevention subjects and teach more advanced and specialized contents related to risk and safety. Furthermore, another advantage is that when workers, fresh from school, are hired in a production company, initially they usually get some kind of safety training, to a higher or lesser degree, to learn the safety basics. This represents a cost for a company that could be eliminated: if the regular education system would teach the safety basics, it would no longer be necessary for a company to do so, and the workers would be able to start immediately without further basic safety education/training, or at least safety training would be a repetition or refreshment of what they already learned in primary- and high school, allowing to tailor the competences to the specific industrial activities of concern. If beneficial, workers could even obtain an advanced and more company-attuned safety education or safety training. In any case, teaching ERM from an early age on represents a win–win situation for society and industry.

2.1.3 Digitalization and education

As the future is moving toward an era of digitalization, ERM education should also deal with this complex subject. Data, analytics, and the digital tools to harness them are transforming all aspects of life, including business and industry. A digital transformation for risk would mean a number of changes. Pivotal among them, risk would capture and manage information from a broader and richer set of data, looking into nontraditional sources. It would automate processes it controls, and work with others to do the same for decision-heavy processes. It would use advanced analytics to further improve the accuracy and consistency of its models, in part by greatly reducing the biases. However, ERM is inherently and appropriately conservative, given its mandate. A change in paradigm is necessary and can only be approached through education.

2.2 Could we make an optimal prevention without education?

It seems easy (and it is tempting) to answer the question posed in the title of this section, with a simple and straightforward "no." Nevertheless, some background information is required to understand the importance of education for prevention. Education and training in general are approaches to gain knowledge, know-how, skills, desired patterns of mind-set and behavior, and the alike, to be able to tackle problems and questions efficiently and effectively, as a person and/or as a group of persons. Furthermore, education and training also create awareness, bring new knowledge, and lead to the further development of adequate skills, attitudes, norms, and values. We cannot say that prevention would be nonexistent without education, but it would certainly not be "optimal."

Safety education in the regular education system, hence to school children, teenagers, and those in early twenties, allows the contents to be very diverse, generalized, theoretical, and not (very strictly) goal oriented. Conversely, education and training in an organizational setting, hence in a company, is usually rather practical, focused, and goal oriented. ERM training in a company should therefore be strongly intertwined with performance management, knowledge management, and the memory of a company.

2.2.1 Educational approaches

Looking at the regular education system, in many countries courses on "traffic safety" are already given to primary school children as well as to high school teenagers (or sometimes called "road safety"). Traffic or road safety lessons for children till the age of 12 seem to be quite effective to teach them to be aware for hazards in the public transportation domain and how to behave in case of specific circumstances or in case of danger. The lessons are usually given in a practical way, and children are taught while "playing" with traffic safety topics. In the teenager category, 12–15 years old, the contents of safety courses may be much more theoretical, and abstract rules and concepts can be taught as well. When teaching to youngsters between about 16 and 24 years old, "theoretical safety" lessons can be added with "experimental safety" lessons: people start learning by investigating the theoretical knowledge and by way of experiments, bringing models and theories into practice. In the working years, ages of about 25 years and older, the focus is placed on learning by doing (practice), learning from case studies, and lessons learned from incidents, added with coaching lessons. In a professional education and training context, when organizational policies, goals, objectives, and activity-based indicators have been determined and communicated to all employees, training and coaching need to be provided to assist them to meet the goals and objectives. In brief and generalized, we can see the following trend:

Play (<12 years old) → **Theory** (12–15 years old) → **Theory** + **Experiment** (16–24 years old) → **Case studies** + **Coaching** + **Practice** (>25 years old)

2.2.2 Person- and behavior-based safety

As Geller [6] indicates, there are a lot of opinions and recommendations on how the psychology of safety can be employed to produce beneficial changes in people and organizations. Most can be classified into one of two basic approaches: *person-* and *behavior-based* approaches. The difference between both types of approaches can be explained by looking at the way how to change people: from the inside (person-based) or from the outside (behavior-based). Person-based approaches attack individual attitudes or thinking processes directly. They teach people new thinking strategies or give them insight into the origin (unsafety) thoughts, attitudes, or feelings. Conversely, behavior-based approaches attack people's behaviors directly. They try to change relationships between behaviors and their consequences, typically providing incentives (rewards or punishments, cfr. the Pavlov experiments in the beginning of the twentieth century).

Person-based safety typically concerns long-term intervention strategies such as education and learning, while behavior-based safety techniques such as manipulation of environmental conditions (or stimuli) are rather short lived, and their effects fade away after maximum about 6 months in case of most people. It is evident that a combination of both types of approaches is optimal for improving safety. Hence, as Geller [6] explains, the person-based approach should be integrated with the behavior-based approach to bring out the best in people and their organizations for the sake of achieving optimal decision-making regarding risks, safety, and performance.

2.2.3 Improved KPABC model

In any organizational context, on the one hand, a lot of information (a.o. about safety) is available and relatively easy to find. However, one has to be open-minded enough to be willing to see the (safety related) information. Such attitude would be the result of the mind-set aspect of ERM education. On the other hand, one has to be able to understand and interpret the information at hand. This requires knowledge, obtained via ERM education. The right mind-set and the right know-how and knowledge for taking optimal decisions in all circumstances cannot be taken for granted. It requires continuous learning, through adequate education and training. Several aspects should be initiated, shaped, sharpened, and/or influenced by adequate ERM education and training: knowledge, perception, attitude, and behavior. All these aspects need to be covered by the ERM education.

Reniers et al. [7] describe the relationship between the aforementioned aspects using their so-called KPABC model. If we combine educational needs as described above with this KPABC model, we obtain an improved KPABC model such as shown in Figure 2.2.

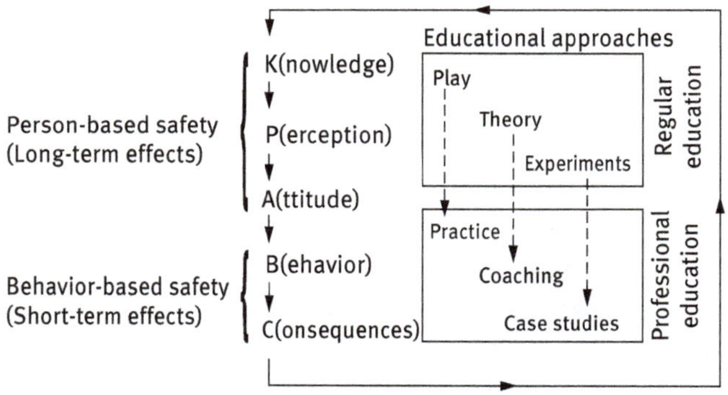

Figure 2.2: Improved KPABC model: Safety education in the framework of the KPABC model. **Source:** After Reniers et al. [7]

A generally accepted method of responding to safety performance problems is to provide education and training. If safety performance levels need to be maintained and even improved on certain domains, then education and training with respect to safety and ERM should be based on specific, measurable, performance-based objectives. Specific topics can, for instance, be related to traffic safety, knowing how to conduct a quick personal risk assessment, first aid competences, reanimation techniques, emergency response actions, and so on.

In light of the improved KPABC model, education and training should be present for every aspect of the model, and thus actual performance management indicators related to education and training should be elaborated for knowledge, perception, attitude, behavior, and consequences. People not only need to learn "safety knowledge," but they also need to be knowledgeable about safety perception, and their safety attitude needs to be in a way that they display safe behavior. One needs to be aware of what the people know, how they perceive safety, what they are able to do about safety issues, what their intended behavior is, and what their actual behavior is.

2.2.4 Safety as a separate education subject – and not

Ideally, ERM, or learning to take the most optimal decisions in personal life and in working life, should both be taught as a separate topic, and also integrated in other courses. The safe way of doing a job is the most efficient way, and if students or employees are taught a good experimental technique, they will have learnt a way which is not only safe but also with the highest performance. Performance can be regarded as the state or the situation of an organization in which as many things as possible go right. This also fully coincides with the newest safety insights on Safety-I and Safety-II. Hollnagel [8] indeed talks about "Safety-I" and "Safety-II." Safety-I indicates all approaches to ensure that as few things as possible go wrong, whereas Safety-II encompasses all approaches to guarantee that as many things as possible go right. Since Safety-I is concerned with things that go wrong or that can go wrong, it is about the ways in which adverse events and adverse outcomes may happen, and about the possible causes and mechanisms that produce the undesired manifestations. Conversely, Safety-II deals with things that go right or that can go right; hence, it is about continuously trying to anticipate developments and events, and making the right decisions leading to excellence and success in the long term. This distinction between Safety-I (safety) and Safety-II (performance) is important, as it leads to some essential mind-set differences of leadership and management. See also Blokland and Reniers [9] for more information on this topic.

For instance, in Safety-I thinking, humans are predominantly seen as a liability or a hazard, whereas in Safety-II thinking, humans are seen as a resource necessary for system flexibility and resilience. Also, on the one hand, in the Safety-I line of thought accidents are caused by failures and malfunctions. The purpose of an investigation in such a case is to pinpoint and determine the causes and the contributory factors. The explanation of accidents in Safety-II language, on the other hand, is that things basically happen in the same way, regardless of the outcome. Thus, the purpose of an incident investigation would be to understand how things generally go right as a basis for explaining how things occasionally go wrong. ERM education is about how to achieve personal and organizational excellence, and as one of the side products, safety excellence.

For example, a chemistry student who is learning how to bore a hole in a cork and fit a glass tube through it, would be taught in his/her normal instruction a safe technique designed to give a nice well-fitting product; as a result he/she should avoid finishing the exercise with a broken end of tubing fitted through his or her hand. Similarly, when learning the techniques of distillation, one should realize that a volatile, flammable solvent should not be heated over a naked flame. Depending on the circumstances and objective, students should, for example, learn not to use a cutting tool with a blunt edge, a hammer with a loose handle, and so on. This sort of information will help to produce a satisfactory product and to avoid personal injuries at the same time. If students, besides gaining specific safety knowledge in a separate course, are regularly brought into contact with procedures and equipment for avoiding any losses, as part of their normal everyday working environment, they will become used to, and accept, safety and ERM through familiarity. The example of professionals who have already been "converted" to an interest in safety and ERM will have a significant positive influence, but continuous learning about safety and risk remains absolutely necessary in organizations. Both regular education and professional education are therefore truly needed to make prevention optimal.

2.3 Standards and education

2.3.1 Standards in the education of engineering risk managers (ERManagers)

It is well known that since more than one century standards are adopted in industrial practices to provide guidance in assuring the quality and compliance of products and production processes. Although in early times standards mostly addressed technical issues and specific technical features of products or of production processes, these were progressively extended to process operation, process and product design, and so on. A number of specific technical standards address operational safety, safety in design, and engineering risk assessment. More recently, standards addressing organizational processes were also issued. In particular, since about three decades, a number of standards concerning safety, health, and environmental management systems in companies were issued by different international and national standardization bodies such as ISO, OHSAS, NORSOK, and HSE. The ISO 45001 standard on "Occupational Health and Safety" and the ISO 31000 standard on "Risk management – Guidelines" represent two examples of the comprehensive and widely applied standards addressing, respectively, the management of safety at work and the management of risk in industrial organizations, available nowadays.

Standards have the merit to embed safety principles derived from high-level expert committees who are involved in issuing and regularly updating the norms. A number of highly valuable principles that should be provided in the education of ERManagers are integrated in standards. Technical standards provide the correct approach to design or operate single components or processes, as well as consolidated safety features of products, and therefore they supply valuable technical data and procedures. Standards addressing organizational and management features embed fundamental educational principles that should be the basis of the competence of an ERManager. Examples of such features include:
- formalization of procedures;
- appropriate documentation and document management;
- appropriate commitment and allocation of resources;
- appropriate planning of processes and procedures;
- appropriate identification of risks;
- appropriate understanding of dynamic features of risk;
- need for continuous improvement and for periodical assessment.

Thus, it is easy to recognize the importance that standards have with respect to ERM.

However, standards neither are intended to substitute competence nor are an educational tool. Rather, due to the huge importance that standards have nowadays in both technical and organizational processes, an ERManager needs to be educated to the use of standards during his or her professional activity. In particular, ERM education should address at least:
- what are standards and how standards are issued;
- what type of standards may be present (international, national, company, etc.);
- what are the obligations concerning the adoption of standards (international rules, national or local legislation, company internal procedures, specific project requirements);
- how existing standards concerning a specific activity may be retrieved;
- how to comply with standards in an organizational process and/or during design or operation of technical systems; and
- which are the limits of standards and how engineering design and operation activities should integrate or be interfaced to standards.

Obviously, professional education should also be completed addressing the content and the requirements of the specific standards applicable in the competence area of interest. However, as stated above, standards are not intended as an educational tool. This means that it is very difficult to learn from standards, or even to fully understand the underlying principles and technical rules in the absence of a strong specific background. Actually, standards are usually conceived as “closed” tools, providing rules or mathematical expressions that need to be applied, with no explanation on the derivation of such rules or expressions, or on the reason for their selection. This should not be seen as a limitation of the standards. In simple terms, they are not intended as learning tools and the discussion concerning the selection of specific approaches or formulas, as well as on their derivation, is usually carried out by the committee that prepares and updates the standard, but is not included in the text of the standard for practical reasons.

Hence, an appropriate application and use of standards in ERM should be based on competent users. Thus by no way it may substitute education on the topics or areas addressed by the standards. Rather, standards may be used to support case studies for coaching or practice that are the last step of professional education for ERManagers, as shown in Figure 2.2.

2.3.2 Competence required by the application of standards in ERM

Since a long time standards were proposed as a solution to complement and integrate the competences of their users in the specific subject area addressed by the standard. However, an open question that is rarely dealt with is the level of competence required for an appropriate use of the standard. It is usually assumed that an appropriate use implies a sufficient level of knowledge by the user, but this is rarely detailed.

In the case of high-level professional profiles, and in particular of auditors who are responsible of assessing the compliance of safety management systems, detailed requirements are stated. This is the case, for example, of ISO 19011 standard [10], providing "Guidelines for Auditing Management Systems." The standard actually includes a detailed list of the competence requirements for auditors, thus defining the level of competence and the background skills required. Educational requirements are, therefore, described in detail, also allowing a more easy access to professional formation provided by specialized companies.

This is not usually the case for nonspecific professional profiles involved at different organizational levels in the risk management process, also at high levels of responsibility.

For example, standard ISO 31000 [11], one of the more widely applied standards addressing "Risk Management Guidelines" for organizations, does not mention any specific requirement on competences. The word "competence" only appears in Section 5.4.4, concerning the allocation of resources, where it is stated that "Top management and oversight bodies, where applicable, should ensure allocation of appropriate resources for risk management, which can include, but are not limited to: people, skills, experience and competence; [. . .]."

As a further example, ISO 45001 [12], addressing "Occupational Health and Safety Management Systems," explicitly recognizes "competence" as a specific support needed for the implementation of a safety management system. The standard states four main requirements concerning competence:

- determine the needed level of competence of the workforce;
- ensure the competence of the workers (by education and training);
- take actions to acquire and maintain the necessary competence; and
- record information that evidences the competence.

Thus, the standard, although recognizing the need and the strong role of competence, leaves to the company the responsibility of determining the appropriate level of competence needed to safely comply with the different roles assigned to the workers.

At the same time, the standard requires that a professional education model, as that shown in the lower part of Figure 2.2, based on coaching and practice, is enforced within the safety management system of the organization.

Actually, this is the key point concerning the educational needs of professional profiles involved in ERM. On the one hand, the need of appropriate academic and professional education is generally recognized and may be a precise requirement of standards. On the other hand, the content of this education is seldom analyzed in detail in standards and/or in official documents. In the case of ERM, the interdisciplinary competences required, and thus the different backgrounds of early professionals that may be called to operate in the field, is a further critical element, since it may be difficult to identify a lack of knowledge in a disciplinary field different from those pertaining to the educational background received.

Moreover, standards are usually conceived as easy-to-use and self-standing documents. Competence required to apply standards is usually an implicit requirement. This may induce a false confidence in the users, leading to the wrong assumption that a very limited competence may be sufficient for their application. Such interpretation is clearly not correct. Actually, standards require a high competence level for their application. This is evidenced in the specific definition of "competence" provided, for example, by standard ISO 45001:

- *competence*: ability to apply knowledge and skills to achieve intended results.

This definition implies, on the one hand, that the persons involved in the management system addressed by the standard have a sufficient background disciplinary knowledge. On the other hand, the definition implies that practical skills allowing the achievement of the intended results of the process are also required. When Figure 2.2 is considered, a clear correspondence of these two requirements to the two steps of the improved KPABC educational model can be recognized: background disciplinary knowledge pertains to the regular and academic education, while practical skills are usually expected to be delivered in professional education. In perspective, the extension of standards providing competence requirements for the different professional profiles involved in ERM may provide a useful contribution to a better framing of educational requirements of ERManagers. Such a process may also contribute to synergies with academic program accreditation requirements, creating a more clear understanding of professional and academic competences required to operate within the different roles of ERM.

2.4 Conclusion

This chapter indicates that educational approaches on ERM differ according to different periods in life, reflecting into so-called regular and professional education. It is argued that without safety education, prevention will never be optimal. Furthermore, to revolutionize safety management in organizations, a focus on five domains, among which safety education holds a prominent place, is needed. If anything, this chapter pleas for making safety as a subject more important in regular and professional education, so that person- and behavior-based safety can both be achieved. With respect to standards, one should be careful not to consider them as a replacement for either education or competence. In fact, to use standards properly, education is actually needed.

References

[1] Swuste, P., Van Gulijk, C., Zwaard, W. (2010) Safety metaphors and theories, a review of the occupational safety literature of the US, UK, and The Netherlands, till the first part of the 20th century. *Safety Science*, 48, 1000–1018.

[2] Swuste, P., Van Gulijk, C., Zwaard, W., Oostendorp, Y. (2014) Occupational safety theories, models and metaphors in the three decades since World War II, in the United States, Britain and the Netherlands: A literature review. *Safety Science*, 62, 16–27.

[3] Swuste, P., Van Gulijk, C., Zwaard, W., Lemkowitz, S., Oostendorp, Y., Groeneweg, J. (2016a) Developments in the safety science domain, in the fields of general and safety management between 1970 and 1979, the year of the near disaster on Three Mile Island, a literature review. *Safety Science*, 86, 10–26.

[4] Swuste, P., Theunissen, J., Schmitz, P., Reniers, G., Blokland, P. (2016b) Process safety indicators: a review of literature. *Journal of Loss Prevention in the Process Industries*, 40 162–173.

[5] Reniers, G.L.L. Khakzad, N. (2017) Revolutionizing Safety and Security in the Chemical and Process Industry: Applying the CHESS concept, *Journal of Integrated Security Science*, 1 (1), 2–15.

[6] Geller, S. (2001) *The Psychology of Safety Handbook*, Boca Raton, FL: Lewis Publishers.

[7] Reniers, G., Ponnet, K., Kempeneers, A. (2014) Higher education chemical lab safety interventions: Efficacious or ineffective?, *Journal of Chemical Health & Safety*, 21(1), 4–8.

[8] Hollnagel, E. (2014) *Safety-I and Safety-II. The Past and Future of Safety Management*, London: Taylor and Francis.

[9] Blokland, P., Reniers, G. (2018) *Safety & Performance. Total Respect Management: A Novel Approach to Achieve Safety and Performance Proactively in Any Organisation*, New York, NY, USA: NOVA publishers.
[10] International Standard ISO 19011 (2018) Guidelines for Auditing Management Systems.
[11] International Standard ISO 31000 (2018) Risk management – Guidelines.
[12] International Standard ISO 45001 (2018) Occupational health and safety management systems – Requirements with guidance for use.

3 Education profiles for ERM

3.1 Different education profiles for ERManagers

A risk manager is required to identify, assess, and manage the risks that affect company businesses and/or assets, and/or that are compromising the safety, security, financial prosperity, and reputation of his/her organization. Within the ERM frame, this requires being able to identify the hazards associated with a technical system or a product, to assess and quantify the risk related to such hazards, to control, mitigate, and manage the risks, and to ensure that the organization is aware and prepared to deal with the potential threats and consequences deriving from such risks. In Section 3.2, a further detail of the required skills and competences is provided.

Hazard identification and risk assessment in the engineering domain require a deep knowledge and a complete understanding of the technical systems in which such risks originate. Even if such activities may not be carried out directly by the risk manager (actually they are more frequently the result of a structured review or formalized assessment activity carried out by multidisciplinary teams), he/she needs sufficient skills to commit the activity to such teams, to assess if the activities were completed satisfactorily and to understand the results.

Moreover, managing technical risks requires affording the issues of prevention, mitigation, and emergency response to each risk identified. More specifically, risk prevention requires dealing with the elimination or reduction of risks by the implementation of inherent safety measures in design, and with the implementation of safety barriers aimed at reducing expected frequencies and the possible impacts of accidents. Engineering risk mitigation also usually relies on the implementation of technical and procedural barriers aimed at controlling and mitigating the consequences and final impacts of the threats related to company activities. Planning of emergency response requires the knowledge and organization of emergency systems and of emergency procedures, and the planning of training activities and of emergency drills. Again, even if all the above activities are rarely carried out directly by the risk managers, ERManagers usually have the responsibility to commit such activities and to assess the results.

The allocation of safety barriers for risk prevention and mitigation, and the deployment of resources for first response and emergency management require the assessment of economic issues related to the implementation of such measures, and appropriate decision-making skills. In particular, quite often, ERManagers are required to deal with "gray" regions in risk assessment, where risks are neither undeniably intolerable nor definitely acceptable. In these decision areas, in some cases defined as "ALARP" (as low as reasonably practicable) or "ALARA" (as low as reasonably achievable) regions, decision-making is usually based on the assessment of costs and benefits of further safety measures. ERManagers thus need to identify,

https://doi.org/10.1515/9783110344578-003

understand, assess, and usually quantify costs and potential advantages of alternative safety barriers. Thus, knowledge of methods for cost–benefit analysis, implying both a strong technical and economic background is an important requirement. More in general, besides the understanding of technical systems, ERManagers need to understand and manage the economic issues related to safety implementation and risk control.

Finally, ERManagers also need to constantly interact with other managerial domains (design, operation, production control, etc.), and are usually responsible for the safety training of operators. Thus, due to the importance recognized to safety climate and safety competences in the safety performance of companies, they need specific social skills to promote safety commitment and to gain a thorough understanding of safety-related issues, even more than managers in other technical disciplines.

Based on the above discussion, ERM requires:

- technical skills, sufficient to understand the managed systems, the hazards and risks deriving from such systems, and the safety barriers available for risk management;
- economic skills, sufficient to carry out economic analyses (e.g., cost–benefit approaches) and other assessments required for decision-making;
- social skills, sufficient to interact with other technical disciplines, to plan safety education and training, and to promote safety commitment and a positive safety climate;
- specific skills, related to the understanding of the state-of-the-art structured techniques for hazard and risk assessment, risk prevention and mitigation, risk monitoring and control.

Figure 3.1 presents an overview of the different fields of knowledge, skills, and activities, required to be understood by an ERManager.

Until the past decade, it was not at all unusual that ERManagers did never receive such specific education, but rather that their specialization was fully based on skills obtained by "learning on the job."

However, nowadays the progress of safety science, the increase in safety requirements, and the growing complexity of the activities required to be carried out by ERManagers make it difficult, if not impossible, to obtain an adequate educational profile by learning on the job. A structured educational process is thus necessary to form ERManagers able to unlock the potential of the state-of-the-art methodologies for risk assessment and management, capitalizing the recent progress of safety science and ERM-related science.

Different educational backgrounds and education profiles may thus be defined for an ERManager. The classic risk manager profile, defined in detail in Section 3.2, is usually based on an educational background that starts with an academic training in basic engineering disciplines (e.g., mechanical engineering, chemical engineering, and civil engineering). Such background is usually obtained by completing at least a B.Sc. or an M.Sc. in one of such disciplines. A few years of experience in operations or design is

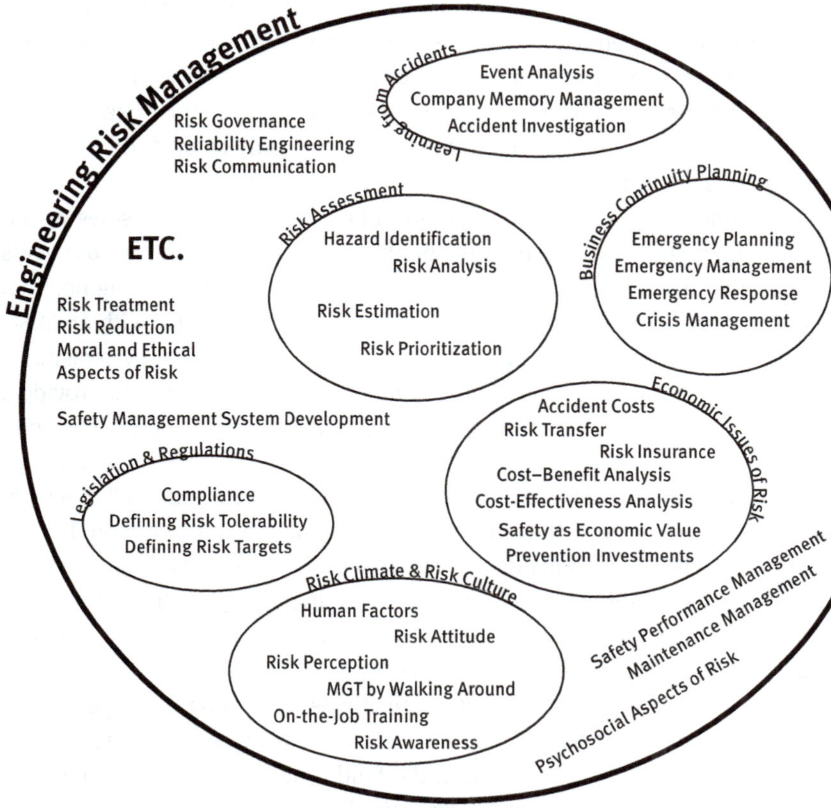

Figure 3.1: A nonexhaustive list of elements belonging to ERM.
Source: Meyer and Reniers [1].

also a common feature of the educational profile of most ERManagers. Both the academic background and the professional experience contribute to warrant a robust understanding of technical systems and of professional practice in operation and design.

The educational profile is then completed by a specific formation addressing the special skills required to risk managers. This is usually obtained by attending professional courses, usually offered by specialized companies. Alternatives are professional master's or specific Ph.D. projects, which a few universities offer worldwide.

A significant issue of this "classic" educational route is that in most universities, B.Sc. and even M.Sc. programs in engineering disciplines do not include safety assessment and risk management as required courses. In most cases, safety is (at least officially) only "embedded" in other technical disciplines (e.g., in design courses). Even if providing safety and risk concepts embedded in technical disciplines may at least be useful to introduce the issue of safety and to provide a basic knowledge of the problem, it is scarcely feasible to deliver a sufficient knowledge of structured methods used in risk management embedding it in technical courses. Therefore, in this classical educational route the exposure to risk management as a discipline and to the structured tools now available in safety assessment comes only late in the professional life. This may be a negative factor, since risk management in this type of educational profile may be intended only as an "add on," or a specific professional skill, pertaining to a design or operational manager.

Alternative educational routes proposed nowadays by several universities worldwide for ERManagers still build on students that have completed a three or four years B.Sc. in engineering disciplines (e.g., mechanical engineering, chemical engineering, and civil engineering). However, rather than continuing the experience with an M.Sc. or professional practice addressing operation and design of technical systems, specific M.Sc. programs addressing safety and risk management are offered to provide a structured ERM background. This route has the advantage of exposing the learner to a well-defined approach to risk management, enforcing the vision that risk management requires a specific professional profile, rather than being an "add-on" to a technical education. As a drawback, this educational profile will result in a more limited knowledge of the technical systems dealt with. However, since nowadays rarely the assessment of technical systems is carried out by a single person, but rather by a team of experts with complementary experience, the higher focus on risk management provided by this educational option is possibly more attractive.

Several universities have started at different times B.Sc. programs specifically aimed at the formation of risk managers. Although in social sciences and economics some of these initiatives were successful, in the field of technical risk management (or "operational" risk management or "engineering" risk management) it is usually a problem to blend competences on technical systems at a sufficient level of detail with a structured knowledge of risk management techniques. Such educational background thus suffers from a rather generic knowledge of the technical problems dealt with in the application of risk prevention and mitigation. Thus, although this option is still open, it is usually not the more frequent educational background of an ERManager nowadays.

3.2 Elements for education of the "classic risk manager profile"

As mentioned by Meyer and Reniers [1], risk assessment connotes a systematic approach to organizing and analyzing scientific knowledge and information for potentially hazardous activities, machines, processes, materials, and others that might pose risks under specified circumstances. The overall objective of risk assessment is to estimate the level of risk associated with adverse effects of one or more unwanted events from one or more hazardous sources. By doing so, it supports the ability to proactively and reactively deal with minor accidents as well as major accidents through "risk management." Hence, risk-based decision-making consists of risk assessment and risk management. The former is the process by which the results of a risk analysis (i.e., risk estimates) are used to make decisions, either through relative ranking of risk reduction strategies or through comparison with risk targets; while the latter consists of the planning, organizing, leading, and controlling an organization's assets and activities in ways that minimize the adverse operational and financial effects of losses upon the organization.

The technical activity of risk assessment, consisting of hazard identification, risk analysis, risk evaluation, and risk prioritization, is thus only one element belonging to the larger domain of risk management. Other elements that are part of risk management are, for example, safety training and education, on-the-job training, management by walking around, emergency response, business continuity planning, risk communication, risk perception, psychosocial aspects of risk, emergency planning, and risk governance. Fuller and Vassie [2] define "risk management" as "the systematic application of management policies, procedures and practices to the tasks of identifying, analyzing, evaluating, treating and monitoring risks."

ERM thus comprises a large variety of activities, knowledge domains, disciplines, and thus learning aspects. Figure 3.1, as mentioned before, shows a nonexhaustive list of elements of ERM.

With the elements of ERM displayed in Figure 3.1 in mind, a distinction needs to be made between learning with respect to a particular task (usually called "training") and learning in general (usually called "education"). Some of the elements mainly need theoretical knowledge (e.g., reliability engineering, risk analysis, and risk insurance), while others require foremost practical and hands-on training (e.g., accident investigation, emergency planning, and management by walking around). All elements, however, need both theory and practice, that is, both education and training.

The task of education is to arrange to supply the right amount of the right kind of information to the learner (student or professional) at the right time. So the question is which parts of which elements of the ERM set need to be taught at which moment in a person's life or career. This may be more difficult than it seems at first sight; Chapter 5 further elaborates this issue.

3.2.1 The process industries as an example

Regarding the elements for education of the classic risk manager profile, a number of different elements are mentioned to be taught in regular education, and a number of other elements are taught in professional training.

A number of different teaching courses and contents for regular education in relation to the process industry, which we take as an example here, are mentioned in literature [3], such as

- Risk, definition, hazards due to technical processes
- Case studies

- Methods for safety indicators of materials, mixtures, and reactions
- Accidental releases and atmospheric dispersion
- Fires
- Explosions
- Methods of hazard assessment
- Methods of risk analysis and risk assessment

Schönbucher et al. [4] mention the following topics:
- Safety and risk management
- Safety assessment of hazardous substances
- Safety assessment of chemical reactions
- Plant safety concepts
- End-of-pipe technologies
- Retention systems
- Process control systems with safety functions
- Accidental releases
- Fire and explosion protection
- Electrostatic hazards

The following safety-related contents are considered as part of the bachelor's program of Safety and Hazard Defence [5]:
- Occupational safety and health
- Introduction to safety science
- Fire protection in structures
- Fire behavior of construction materials and elements
- Explosion protection
- Chemistry of fires and extinguishing
- Safety indicators
- Technical risks
- Dispersion of hazardous materials
- Safety concepts
- Ecology
- Psychology for emergency responders
- Legal issues in fire and disaster response
- Operational management in hazard defense
- Security of estates

In the industrial safety specialization of the master's program of Safety and Hazard Defence, the following courses are addressed:
- Mathematical modeling
- Probabilistic safety analysis

- Heat engineering
- Control of exothermic reactions
- Safety characteristics
- Shutdown management
- Dynamics of complex flows
- Legal issues of industrial safety
- Legal issues of occupational safety
- Information and communication technology
- Safety laboratory exercises
- Special subjects of safety research

Jönsson and Lundin [6] mention the following contents given to students of the Master of Science in Risk Management and Safety Engineering:
- Statistical methods for safety analysis
- Managerial economics
- Risk analysis methods
- Safety, health, and environmental law
- Economic models for risk management
- Risk management processes
- People, technology, organization, and risk management

These are just examples of contents and courses of programs that have been set up for the classic risk manager formed in regular education. It is clear that the focus is heavily put on theoretical knowledge about technology and hard science knowledge, law, economics, mathematics and statistics, phenomena of processes that are relevant for technical safety (e.g., runaway reactions, fires, and explosions), risk assessment techniques and approaches, and so on.

In professional education, also often referred to as “training,” much more emphasis is put on applied knowledge and know-how. Lees [7] mentions the following contents to be taught to operators in the process industry, as part of classic risk management:
- Process goals, economics, constraints, and priorities
- Process flow diagrams
- Unit operations
- Process reactions, thermal effects
- Control systems
- Process materials quality, yields
- Process effluents and wastes
- Plant equipment
- Instrumentation
- Equipment identification

- Equipment manipulation
- Operating procedures
- Equipment maintenance and cleaning
- Use of tools
- Permit systems
- Equipment failure, services failure
- Fault administration
  - Alarm monitoring
  - Fault diagnosis
  - Malfunction detection
- Emergency procedures
- Fire fighting
- Malpractices
- Communications, record-keeping, reporting

King [8] mentions the following safety-related contents to be taught in organizations to managers:
- Managerial responsibility for safety and loss prevention
- Legal requirements
- Principles of safety and loss prevention
- Company safety policy, organization, and arrangements
- Hazards of the particular chemicals and processes the company uses
- Accidents and accident prevention, statistics, and case studies
- Pressure systems
- Trip systems
- Principle of independent assessment
- Plant maintenance and modification procedures, including permits-to-work and authorization of modifications
  - Fire prevention and protection
  - Emergency planning arrangements
  - Training of personnel
  - Information feedback
  - Good housekeeping
  - Sources of information on safety and loss prevention including both people and literature
  - Case histories

Furthermore, the professional course by Wise Global Training [9] for instance treats the following contents:
- Learning from incidents
- Hazards inherent in oil and gas

- Risk management techniques used in the oil and gas industries
- An organization's documented evidence to provide a convincing and valid argument that a system is adequately safe
- Contract management
- Process safety management
- Role and purpose of a permit-to-work system
- Key principles of safe shift handover
- Plant operations and maintenance
- Start-up and shutdown
- Failure modes and other types of failure
- Safety critical equipment controls
- Safe containment of hydrocarbons
- Fire hazards, risks, and controls
- Furnace and boiler operations
- Fire and explosion in the oil and gas industries
- Emergency response
- Marine transport
- Land transport

The Institution of Chemical Engineers in UK offers in 2018 a professional course in process safety management that identifies the following six pillars:
- Knowledge and competence
- Systems and procedures
- Human factors
- Engineering and design
- Assurance
- Culture

It is important to mention that all the above lists of topics, either intended within regular university programs or as part of professional courses, include both specific safety- or risk-related items (e.g., hazard assessment techniques, risk assessment, and risk management techniques) and technical concepts (e.g., unit operations, control systems, pressure systems, and trip systems).

3.2.2 The classic risk manager profile

It becomes clear from the previous section that the traditional way to form risk managers or safety-caring employees is to provide them with theoretical courses during their regular education, and to provide practical courses, coaching, and on-the-job training in their professional life. This is also what was shown in Fig. 2.2 displaying the improved KPABC model.

Looking at the contents of the courses, we can conclude that the education is rather focused on very specific knowledge domains (Knowledge, gained via regular education) or on carrying out very detailed tasks (Behavior, gained via professional education), and largely "forgets" educating the other aspects of the improved KPABC model, that is, Perception and Attitude. Ideally, all aspects should nevertheless be taught in an integrated and multidisciplinary manner. Chapter 5 will elaborate further on this.

3.3 Elements of education of a continuing education profile

What people learn in their youth will remain valid and useful for the rest of their lives. Knowledge gained at any point of time is largely obsolete within a matter of years. Skills that made people productive in their twenties become out-of-date in their thirties [10]. So it is no longer functional to define education as a process of transmitting what is known: it must now be defined as a lifelong process of continuing inquiry. Thus, the most important learning of all is how to learn the skills of self-directed inquiry. People become ready to learn something when they experience a need to learn it in order to cope more satisfyingly with real-life tasks or problems.

Especially in the risk management field, the educator has a responsibility to create conditions and provide tools and procedures for helping learners discover their "needs to know." Learning programs should be organized around life-application categories and sequenced according to the learners' readiness to learn. As people grow and develop they accumulate an increasing reservoir of experience that becomes an increasingly rich resource for learning-for themselves and for others. We often hear that management can improve with experience. This is also the case when dealing with risks; however, experience by itself will not bring a sufficient added value. Furthermore, people attach more meaning to learning gained from experience rather than from passive learning. Accordingly, the primary techniques in education are experiential techniques – laboratory experiments, discussion, problem-solving cases, simulation exercises, field experience, and the like. Learning experiences should be organized around competency-development categories. People are performance centered in their orientation to learning.

Understanding the personal and professional motivations that shape learning needs is fundamental to high-quality people working in any kind of activities. Continuing education could serve as the facilitator of dialogue, research, and policy to address risk management issues. The domain of continuing education encompasses also performance appraisal, personal training, teacher development from the business and professional community, and others. The majority of continuing education programs tend to be focused on structured and planned programs as opposed to self-directed learning. Determining an appropriate learning process is key to implementing continuing education successfully, but it is equally important to measure if continuing education programs or self-directed learning have achieved their goal.

Traditionally, each level of a complex system is studied separately by a particular academic discipline, and modeling is done by generalizing across systems and their particular hazard sources. Risk management must be modeled by cross-disciplinary studies, considering risk management to be a control problem and serving to represent the control structure involving all levels of society for each particular hazard category.

The usual approach to modeling sociotechnical systems is by decomposition into elements that are modeled separately (analytical system). An additional approach for ERM should be based on a systemic approach analyzing the relationship among all the elements and their environment. This has some peculiar effects. The sociotechnical system involved in risk management is normally decomposed according to organizational levels, which are then the objects of study within different disciplines. The effect of this is that risk management at the upper levels is normally studied with a "horizontal" orientation of research across the technological hazard sources. Traditionally, sociological studies at the upper levels are based on the analysis of samples of organizations or groups of people with no detailed consideration of the actual processes found at the productive bottom level [11].

Continuing education could be realized at different stages: the first is a tutor–mentor process where a more experienced person trains, at the workplace, a less experienced one. This might be the most effective one at a first glance, but highly dependent on the goodwill of the tutor/mentor and, of course, of the learner. Tutoring is more a relationship of a fixed duration, where the tutor is knowledgeable about the subject area and able to pass on skills and knowledge. It focuses on learning specific skills and knowledge. Mentoring is rather an ongoing relationship, where informal meetings take place when the "mentee" needs some advice, guidance, or support. Within the same profession and/or organization, mentors are usually more experienced and qualified than "mentees" and can, therefore, pass on experience and knowledge. Generally speaking, it focuses on career and personal development of the "mentee." The qualities that are essential in a mentor may differ depending on the disciplines. For the engineering domain, Meyer proposed [12]:

- Willing to share skills, knowledge, and expertise
- Being available for the mentee
- Being prepared beforehand (technical, organizational, personal, resources)
- Provides guidance and constructive feedback
- Values the opinions and initiatives of others
- Motivates others by setting a good example
- Demonstrates a positive attitude and acts as a positive role model
- Takes a personal interest in the mentoring relationship

- Retain objectivity and fairness and does not unfairly influence any process either of one may be involved in
- Being approachable, flexible, and open-minded
- Last but not least . . .. have fun

The second stage is to enforce people to get through compulsory trainings or courses depending on their activities and hazards they will face. This is quite a normal process in industries and a legal concept in most countries. However, in some sectors, for instance in academia, the term "compulsory" is felt like an attack of the "holy (e.g., academic) freedom." This form of training should be reserved to specific fields where safety must be ensured and no compromise could can be made about the effectiveness of the education's result.

The third level is voluntary enrollment in optional safety courses. This continuing education may also be proposed to professionals outside the institution giving the course. They encounter generally a great success despite the fact that they are not free of charge (for people outside the institution giving the education).

The fourth stage is to incorporate pre-course reading, informal and interactive seminars, practical demonstrations, and role-play during scenarios. It is important that participants undertake role interchange during scenarios, thereby facilitating mutual understanding. They should be encouraged to reflect on their actions and to pay particular attention to detail.

The last stage is educational meetings, which are frequently combined with audit and feedback as well as educational outreach.

Finally, an evaluation process should be developed to determine if continuing education is achieving its desired outcome. One way to determine how much time professionals should spend on continuing education is to obtain some form of measure that indicates when the goal of continuing education has been attained. Survey, exams, inquiries, observational studies, monitoring improvements, data mining and analysis, and so on are some possible measurement techniques. How to measure the efficiency of education is a topic by itself and will not be debated here.

As mentioned by Leveson [13] in her book, engineers are struggling to apply old techniques to new software-intensive systems, expending much energy and having little success. At the same time, they can no longer focus only on technical issues and ignore the social, managerial, and even political factors that impact safety if they are to significantly reduce losses. The solution is to build on the unique approach to engineering for safety coming from the education.

Although continuing education applied to risk management remains largely didactic at present, the importance of moving to a participatory paradigm has been noted in a systematic review by O'Brien et al. [14]. They considered the effects of educational meetings, such as lectures, workshops, and courses, on professional practice and healthcare outcomes. These authors concluded that interactive workshops, rather than lectures and academic instruction, were the best way to influence changes in professional practice. The next section further expands on this observation.

3.4 Different forms of safety education

Research indicates that people learn through the five senses, and the more senses are used to learn, the higher the learning result and its impact. Handley [15] explains that people learn through seeing 80% of the time. Hearing serves to learn 14% of the time, while the three remaining senses (touch, taste, and smell) each individually account for learning 2% of the time. Since the spoken word is very often misunderstood or misinterpreted, it is not an optimal means to communicate a message or to teach. Most experts agree that there is a 30% more comprehension and probably 50% more retention when multisensory channels of communication are employed.

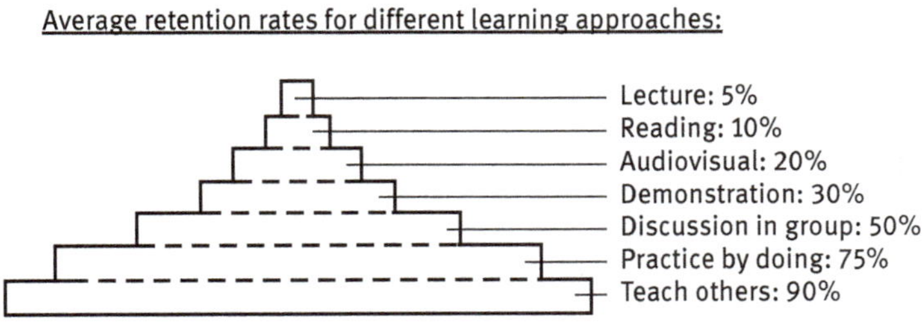

Figure 3.2: The retention pyramid.
Source: Roughton and Mercurio [16].

If we look at how people remember what is educated, hence how an individual learns, studies indicate that giving a lecture actually shows the lowest retention rate. Figure 3.2 provides some learning approaches and their retention rates.

From Figure 3.2, it follows that trainees mainly learn through practice. Of course, this is especially the case when people are learning hands-on skills. Conversely, abstract concepts or theories are much more difficult to be taught via "practice by doing," so different learning methods should be used to this end. Different forms of safety education are indeed available for trying to achieve high retention rates.

There are four types of education depending on the location where the education takes place:
(i) Learning in the classroom
(ii) Electronic learning (e-learning)
(iii) Learning in the field
(iv) On-the-job learning

For each of these different types of education, a variety of possible methods can be used. An overview can be found hereafter and are explained more in depth in the following paragraphs.

Learning in the classroom:
- Lecturing
- Discussion
- Panel discussion
- Workshop/theoretical exercises
- Case study
- Role-playing
- Laboratory training
- Demonstration
- Showing films
- Simulation
- In-basket simulation
- Flipped classroom

Electronic learning:
- Long-distance learning
- Massive open online course
- Blended learning

Learning in the field:
- Field visits
- Observations of practice
- Safety audits

On-the-job training:
- Job-instruction training
- Internship or assistantship training
- Job rotation
- Coaching

Hereafter, the different existing approaches are elaborated.

(i) *Learning in the classroom*

Most educational activities probably find place in a classroom context. Especially in regular education, the classroom is a well-established surrounding for learning activities. But also in industrial practice, many times people learn in the classroom. There is a variety of different methods that can be employed in a classroom, which will be discussed hereafter.

The most used method in the classroom is by no doubt *lecturing*. Major advantages of this method are that it is relatively cheap and that factual information can be given to relatively large numbers of people at the same time. This benefit of the method, however, comes with a downside: mere exposure to information in no way guarantees learning. Figure 3.2 indicates that the average retention rate is only 5%. The rate can be augmented by making the lecture more interactive and by avoiding one-way communication: creating opportunity during the lecture to clarify meanings, checking whether the learners really understand the lecture material, handling the wide diversity of ability, attitude, interest that may prevail among the trainees, and others. As explained by Setty and Moorthy [17], the competent lecturer should make the material meaningful and intrinsically motivating to his/her listeners.

Another method for the classroom is the *discussion*, sometimes also called "conference". A discussion can be employed for a variety of objectives, among others providing information, collecting opinions and finding facts, exchanging opinions, decision-making (or problem solving), and decision "selling." According to Setty and Moorthy [17], this method is particularly interesting for acquiring and understanding conceptual data and for the development or modification of attitudes. Some advantages of the discussion is that it facilitates different viewpoints, it is by nature participative and interactive, it allows immediate feedback, it leads to a "group" feeling of the learners, there is direct and indirect learning, and so on. A disadvantage is that it is not possible to use the method for large numbers of learners, although there are ways to solve this problem, for example, by forming small(er) groups and having a discussion via groups' spokesmen. Another possible downside (at least for some purposes) is that the discussion technique often requires some maturity of learners. Nonetheless, if it is possible to use the discussion method, as indicated in Figure 3.2, the average retention rate is high, some 50%. The increased learning follows from the fact that, instead of presenting all the answers, questions are posed and conditions are created to enable the learners to discover answers for themselves and then evaluating them.

A particular type of discussion is a *panel discussion*. In this case, some four or five experts are invited to sit in a panel before a wider audience, and each of them will be focusing on a particular dimension of a topic or issue and he/she will share his/her views and experiences. There is also a moderator to lead the panel discussion. Initially the panelists discuss among themselves, while the audience will be listening to the viewpoints and arguments as expressed by the experts. Afterwards, the discussion is usually thrown open to the audience, and learners from the audience may pose questions or seek clarifications from the individual panelists. An advantage of this type of discussion is that people from different walks of life, with experience and depth in their own areas of specialization, express their views, opinions, and objective information, adding to the knowledge of the learners.

Yet another method for the classroom is the *workshop* (sometimes also called '*theoretical exercises*'). Instead of directly teaching or instructing the learners, they are involved – individually or in groups – in working out an exercise or an aspect related to a subject or an issue. At first, the whole group of learners need to be familiarized with the theory behind the subject or issue, for example, by way of lecture, and eventually also by sample exercises. Afterward, the learners will either individually or in small groups elaborate, solve, and work out one or more exercises themselves. Essentially, in this method the trainees learn by doing in a theoretical way, or some kind of self-learning. In this sense, there may be classroom workshops/exercises, but also home-based workshops/exercises. Receiving individual feedback on the quality of the learners' workshop/exercise results is important for assuring learning.

The *case study* is another method that can be used in the classroom. This method is based on the belief that understanding and competences can best be attained through the study, contemplation, and discussion of very concrete cases. The learner is presented with a written case that describes a concrete problem. He/she is usually asked to study it in a group, in order to discuss the problem-solving approach related to the concrete case and trying to derive the principles involved. Since no single correct solution is applicable to most cases, learners are encouraged via case studies to develop flexibility in their approach to solutions finding. The case study also helps drawing inferences, motives, thinking and behavior of learners, and the way learners act or react. This way, trainees not only learn to capture the issues involved in a concrete situation (whether policy, conflict, decision-making, etc.) but also they learn to understand their views, attitudes, biases, prejudices, and so on. Hence, the process of working through a concrete case with other learners in a "free" discussion environment provides a "checks-and-balances" approach and leads to learning to deal with relationships and elements in a situation with a minimum of personal bias. Cases may even vary from very brief summaries of small problems to 70-page detailed descriptions of problems. Since case studies can be regarded as a combination of "discussion in group," "practice by doing," and "teach others," the average retention rate may be estimated at around 70%.

Role-playing is a method which is related to case studies in the sense that it requires a very concrete problem/setting. It involves participants' assuming specified roles and acting out certain events and/or facts. Hence, instead of talking and discussing about solving problems, learners need to spontaneously "play out" solutions to these problems as they think that the persons whose role they are playing would or might view them. Setty and Moorthy [17] indicate that role-playing has proven to be an effective technique for promoting real attitude change. Studies indicate that, regardless of whether the player is satisfied with his/her performance, he or she is being forced to improvise a defense for a certain role he/she takes, and this way the likelihood that he/she will end up convincing himself or herself of the validity of this different viewpoint increases. Several flaws can be mentioned for this technique: learners may find it rather childish, so they should be convinced of the importance of this specific learning experience, or some players are likely to overdramatize and overact, whereby they put more energy and focus on the acting than on problem solving. The average retention rate of this approach depends on the context in which the role-playing takes place and whether the case material for the role-playing was well prepared by the teacher and whether adequate discussions were held after the role-play.

Another technique that may be employed in the classroom, is *laboratory training*. This approach serves to learn by doing, be it in an experimental setting, and as such it is "experience-based learning." The approach enables learners to acquire a realistic appraisal of what they can and cannot carry out and achieve in experiments, and what safety rules should be considered and why.

A technique related to laboratory training is *demonstration*, thus showing how a particular task is to be carried out. This technique facilitates in teaching a certain task, introducing a new procedure, or illustrating a certain point. It mainly deals with showing what to do, how to do it, and when to do what.

Another learning method for the classroom and related to demonstration is *showing films*. The advantage of films is that they provide at the same time audio and visual aspects of a topic. Moreover, films are usually structured and entertaining and, once developed, they are suitable for a wide range of subjects and audiences. A downside of films is that they may cover only partially the subject at hand and that they are not interactive at all. Usually, films are used in combination with another method such as lecturing or case study.

A learning technique requiring highly technological classrooms is *simulation*. Examples can readily be found in aviation or in space exploration. Airplane pilots and astronauts need to go through many hours of simulation training in various circumstances that simulate certain conditions and/or situations. Other industrial sectors such as the process industries are also using simulation as a technique, but it could certainly be employed on a much wider scale for learning related to high impact low probability events, also called type II events. Recently, 3D simulations based on laser scanning to reproduce complex working environments, related to a chemical or power plant, were introduced for safety training. Such simulations, however, require very expensive and highly complex infrastructures.

The so-called *in-basket simulation* is a particular form of simulation that can easily be used, thanks to its lack of complexity. The learner is given a set of written materials, letters, information of different kinds, containing background details of (a) certain problem(s). The student then needs to solve the problem(s). The amount of information available for making decisions is usually limited (there may also be given superfluous information), as is the time allowed for completing the simulation process. In-basket simulation is very good at testing learners' antistress capabilities.

Another type of learning in the classroom is called *the flipped classroom*. This type of blended learning reverses the traditional learning environment by delivering instructional content, often online, outside of the classroom. In a flipped classroom, students then watch online lectures, collaborate in online discussions, or carry out research at home while engaging in concepts in the classroom with the guidance of a mentor. In brief, this approach moves activities, including those that have traditionally been considered "homework," into the classroom.

(ii) *Electronic learning (e-learning)*
Distance education or *long-distance learning* is the education of students who are, for one reason or another, not able to be physically present in a classroom. Traditionally, this usually involved correspondence courses wherein the student corresponded with the school via post. Today it involves online education or e-learning. Massive open online courses, offering large-scale interactive participation and open access through the World Wide Web or other network technologies, are recent developments in distance education.

Indeed, a popular method for learning electronically is to follow a massive open online course, better known by its abbreviation '*MOOC*'. MOOCs are available on a variety of topics, and anyone can subscribe to such a course. It is an online course aimed at unlimited student participation and open access via the World Wide Web. In addition to traditional course materials such as filmed lectures, readings, and problem sets, many MOOCs provide interactive user forums to support community interactions among students and teachers. MOOCs are a recent and widely researched development in distance education, which were first introduced in 2006, emerging as a popular mode of learning in 2012 and still gaining in popularity worldwide. It is possible to develop MOOCs for ERM purposes.

Another method for e-learning is called *blended learning*. Blended learning can be regarded as an education program that combines online digital media with traditional classroom methods. It requires the physical presence of both teacher and learners, with some elements of student control over time, place, path, or pace. While learners still attend traditional classrooms with a teacher present, face-to-face classroom practices are combined with computer-mediated activities regarding content and delivery. Blended learning is also used in professional development and training settings. By using a combination of digital instruction and one-on-one face time, learners can work on their own with new concepts, which frees teachers up to circulate and support individual learners who may need individualized attention. An advantage of blended learning is that teachers can attune their instruction to help all learners to reach their full potential. The use of information and communication technologies has been found to improve student attitudes toward learning, and the average retention rate is expected to be higher than while using a traditional lecture education approach.

(iii) *Learning in the field*
Learning in the field can be achieved via so-called *field visits*. Field visits are carefully arranged group visits to an industrial area, a certain place (e.g., an innovative

company or university-related laboratory), or an institution for onsite observation and learning. They require planning (among others the scope of the visit), coordination, and analysis of the messages and information/data that were given at the visit, or at the end of the visit. The goal of such visits is to observe ongoing activities at a specific location, to verify, compare, and confirm with respect to theory, issues, problems, solutions, means, and so on. Through field visits, learners learn by seeing, but they also learn by asking questions during the visit to the local organizer, and they learn by interacting and exchanging ideas and views with fellow learners. Practical demonstrations can also be organized during the visit, which are difficult to prepare in a conventional classroom. As such, the average retention rate is higher than merely that of an audiovisual learning method.

A method linked to the “field visits” approach for learning in the field is *Observations of practice*, where visits are organized to specific practical tasks carried out at an organization or in an industrial surrounding. The observation-learners are individuals or very small groups (two to three persons) that observe without asking questions or interacting with the persons carrying out the tasks. An example is observation-learners at a Hazop (Hazop is a much used and widely known risk analysis method in the chemical and process industries) exercise: the trainees learn by carefully observing how the Hazop is carried out. This is a combination of the “audiovisual” and “demonstration” methods of learning.

Yet another method for learning in the field is the carrying out of *Safety audits*. This method is mainly suitable for learners who are already relatively experienced in safety. A safety audit examines and assesses in detail the standards of all facets of a particular activity. It extends from complex technical operations and emergency procedures to job descriptions, house-keeping, and attitudes. Typically, an audit has a connotation that one needs to find something wrong. We need to get over this attitude and stop looking for all the “bad” things, but consider safety audits as a means of “conformance appraisals”, where an auditor looks at how expectations are met and focusing on the positive aspects of safety. A safety audit needs to be well prepared by the auditor (in case a learner), and at the location photos need to be made as evidence of audit conclusions and recommendations. It is a form of learning by teaching others, and therefore the average retention rate of this learning approach is very high.

(iv) *On-the-job learning*

The idea behind on-the-job learning is surprisingly simple and practical: employees learn while they produce and have earnings. Since no specific space/location and/or equipment are required for this method, it is probably the most common way that employees receive instruction in organizational settings. An important advantage of on-the-job learning is also that learners are taught in the physical and social environment in which they are already working, or will eventually perform their job.

A disadvantage of on-the-job learning is that expensive equipment and workspace are tied up during the learning. Moreover, though in theory it is possible to create maximally favorable conditions for learning on the job, in practice the workspace has a primary production function, and not a learning function, and as such circumstances will never be "ideal" for the learning.

The first type of on-the-job learning is the so-called *job-instruction training* (JIT). The first part of JIT is to formally introduce an employee to his/her specific job environment, to the people with whom he/she will be working, and to organization policies and objectives that concern him/her. The second part of JIT is to provide the employee with a step-by-step review and demonstration of the job tasks and operations. The employee-learner then tries to perform the tasks himself/herself, explaining as he/she proceeds. This process continues and is repeated until both the instructor and the employee-learner are confident that the latter can perform the necessary operations or tasks on his/her own. A downside of JIT is that it is done under normal operating conditions, and thus no specific situation or conditions are considered. The average retention rate is expected to be rather high since the learning method is a combination of demonstration, practice by doing, and teaching others.

Another approach for on-the-job learning is the *Internship or Assistantship training*. In this case, the trainee is assigned a temporary job at an organization, to be engaged in one or several assignments to familiarize him/her with the many problems encountered at the workplace of this organization. Over a certain period of time, the trainee's supervisor observes his/her performance and reviews his/her shortcomings and strong points, thus helping him/her to improve his/her performance on future assignments or tasks. During an internship, a trainee learns by doing.

Job rotation, a third type of on-the-job learning, provides the employee (-trainee) with a variety of assignments and tasks from within the organization. Well-planned job rotation leads for the employee(-trainee) to acquiring valuable perspective and knowledge of interrelationships while experiencing the problems and approaches of various organizational functions. An advantage to the company may be the cross-fertilization between different organizational departments. In essence, job rotation is similar to an internship training but longer term and more systematic education via learning by doing.

A last method for on-the-job learning is popular in many companies and can be summarized as *coaching*. This is a flexible and direct method of learning, which allows the trainee to develop and apply his/her skills under the guidance of a skilled coach who has the time, patience, and desire to help trainees. The coach keeps a careful record of the progress and attitude of each of his/her trainees. Because of its personal nature, the learning method can be very effective. The coach can be someone belonging to the company, or he/she can be a consultant.

3.5 Conclusion

ERM does not take place in isolation; drawing out different perspectives, developing a commonality of understanding, as well as determining clear roles and responsibilities for managing them are all facets of an effective approach. By having all the key stakeholders round the table undertaking an open discussion of the risks, one can be more assured that effective ERM will take place. Education will play a key role within this process.

Traditional risk areas such as employee liability, rising operational costs, and safety and security concerns are growing in complexity while new areas such as cyber liability, duty of care requirements, and bedbugs add to the perceived burden and cost of risk management. In short, institutions need to do more with less, to develop innovative, cost-effective risk management solutions that address the unique requirements of higher and interdisciplinary education needs.

Like many professions, those who lead an organization's engineering risk management efforts often develop their own language that they use to communicate with others. Conversations about likelihood, impact, inherent and residual risks, risk appetite, and risk tolerances become commonplace among engineering risk managers, but they may not be well understood by others in the business or at the governance level. Some ERManagers forget the importance of speaking the language used by those in their audience. Managers are focused on boosting margins, achieving objectives and goals, and advancing the business and that usually affects how they think and the language they use. As a result, the language ERM professionals tend to use may not be well understood or appreciated by senior or experienced managers. New teaching in ERM should also tackle the problematic of communication, not solely with external but also internally concerning the institution or the company.

A global strategy aims to respond to the needs of society and industry concerning operational risks by improving the education on ERM in the graduate and postgraduate courses, developing the human resources with practical knowledge and skills for ERM in different sectors, providing training opportunities for business workers regarding ERM, and promoting public outreach on ERM.

References

[1] Meyer, T., Reniers, G. (2016) *Engineering Risk Management*, Berlin: De Gruyter.

[2] Fuller, C.W., Vassie, L.H. (2004) *Health and Safety Management: Principles and Best Practice*, Harlow, Sussex, UK: Pearson.

[3] Krause, U. (2016) Process safety in engineering education – pro's and con's of different approaches, *Chemical Engineering Transactions*, 48, 871–876.

[4] Schönbucher, A., Brenig, H.W., Hauptmanns, U., Klais, O., Schmidt, J., Moritz, H.U. (2012) Lehrprofil Prozess- und Anlagensicherheit (Teaching Profile Process and Plant Safety, in

German), Dechema – Gesellschaft für Chemische Technologie und Biotechnologie e.V., Frankfurt/Main, Germany.

[5] OvGU (2013) Otto von Guericke University, Module Handbook "Sicherheit und Gefarhrenabwehr" (Safety and Hazard Defence).

[6] Jönsson, R., Lundin, J. (2001) Master of Science in risk management and safety engineering, at Lund University, Sweden, In: *10th International Symposium of Loss Prevention and Safety Promotion in the Process Industries*, Pasman, H., Fredholm, O., Jacobsson, A. (Eds.), Amsterdam: Elsevier.

[7] Lees, F.P. (1996) *Loss Prevention in the Process Industries*, 2nd Ed., Oxford, UK: Butterworth-Heinemann.

[8] King, R., Hirst, R. (1998) *King's Safety in the Process Industries*, 2nd Ed., London, UK: Arnold.

[9] Wise Global Training Ltd., *Introduction to Oil and Gas Operational Safety*. London: Taylor and Francis, 2015.

[10] Knowles, M.S. (1970) *The Modern Practice of Adult Education*. Prentice Hall Regents: Cambridge Adult Education.

[11] Rasmussen, J. (1997) Risk management in a dynamic society: a modelling problem, *Safety Science*, 27 (2–3), 183–213.

[12] Meyer, Th. (2017) Towards the implementation of a safety education program in a teaching and research institution, *Education for Chemical Engineers*, 18, 2–10.

[13] Leveson, N.G. (2011) *Engineering a Safer World: System Thinking Applied to Safety*. Cambridge, USA: MIT Press.

[14] O'Brien, T., Freemantle, N., Oxman, A.D., Wolf, F., Davies, D.A., Herrin, J. (2003) Continuing education meetings and workshops: effects on professional practice and health care outcomes. *The Cochrane Library* (2), Oxford.

[15] Handley, W. (1977) *Industrial Safety Handbook*. Maidenhead, Berkshire, England: McGraw Hill.

[16] Roughton, J.E., Mercurio, J.J. (2002) *Developing an Effective Safety Culture*. Oxford: Butterworth Heinemann.

[17] Setty E.D., Mourthy, P.K. (2008) *Training Methods Focusing on Industrial Safety*. New Delhi, India: Akansha Publishing House Ltd..

4 Learning characteristics and outcomes

4.1 Safety teaching and training

As evident from the previous chapters, in the present society, safety teaching and training and learning about safety is essential for every profession and in every organizational context and industrial sector. Safety education and learning about safety is a process consisting of three aspects (see also Figure 2.2 for the (bigger) conceptual picture):

1. Imparting knowledge and information (to have sufficient safety erudicity)
2. Creating awareness and the right attitude (to attain the correct perception)
3. Developing skills and know-how (to know what to do and how to behave)

If one of those three aspects are not adequately taught, safety learning will not be optimal. The learning characteristics and outcomes should thus take these three aspects of safety education into consideration. Education to safety knowledge, a safety attitude, and safety skills are truly necessary if an organization desires to make profits by gaining hypothetical benefits (cf. also [1, 2]) and be profitable in the long term. Similar to a soldier or a general in the army who needs to be taught knowledge, attitude, and skills about their role in warfare, an employee or a (top-) manager indeed needs to be taught safety knowledge, attitude, and skills for adequate management leading to a profitable company. If you have the right knowledge to perform a task, but you lack the skills and/or the attitude, the knowledge serves for nothing. Analogously, if you have the right skills, but you lack knowledge and/or the proper attitude, safety is not guaranteed while performing a task. Evidently, if one has the right attitude and is positively inclined to carry out a task taking safety very seriously, but he/she lacks knowledge and/or skills, task safety will not be ensured.

One of the most visible learning characteristics about safety concerns the diversity of safety areas, both from a disciplinary viewpoint (safety science, health sciences, law, toxicology, chemistry, mathematics, physics, biology, management and economics, etc.) and from an area/field/safety element perspective (risk assessment, safety economic aspects, business continuity planning, regulation, human and organizational factors, etc.; see also Figure 3.1). Safety characteristics and outcomes further depend on specificities within various industrial sectors such as the manufacturing industry, the chemical and process industry, the construction industry, the space industry, fishery, the pharmaceutical industry, the food industry, and any other you may consider.

https://doi.org/10.1515/9783110344578-004

It should be noted that safety education is most effective when it is aimed at defining needs, that is, performance management outcomes, in order to improve personal knowledge, or to contrast a lack of skills and/or a lack of attitude. It is, however, a waste of time when safety needs or safety problems are situated elsewhere in the organization, and/or there are system issues.

Safety education starts with one essential step, that is to define the goals and the objectives of the education. As Petersen [3] indicates, safety objectives should always be determined and clear before content and methods for education are decided upon. As is clear from the previous chapter, other methods than telling (lecturing) and showing (on-the-job training) are possible. The objectives can, for example, be developed and fixed in terms of desired behavior. For example, the following safety-related goals at the end of a course for managers can be envisioned: (i) transmit organizational safety policies to employees, (ii) orient each new employee in safety, and (iii) make sure that every employee carries out a job safety analysis before starting his/her job.

But who should be educated? In regular education, ideally all students in primary school, high school, and higher education should be taught some basics of safety and preventive decision-making. In professional education, everyone in an organizational context should enjoy safety education on a regular basis: new employees, experienced workers, any employee transferring to another job, and groups of employees.

Setty and Moorthy [4] suggest three types of safety courses:

1. Courses covering substantive technical areas, knowledge, and skills for professional growth
2. Courses for personal growth and efficacy
3. Courses for interpersonal relations and managerial skills

These three types of courses, given throughout a person's life and career, should subsequently lead to higher personal achievements as well as meeting organizational goals and objectives, from both safety and performance viewpoints. Besides the content for the courses being derived from applied technological sciences, exact sciences, and human sciences, there should also be content such as how to motivate people, communication skills, leadership, personality development, management of change and conflict, training methods and skills, how to carry out good "performance appraisals," team building, and the alike.

The idea of the courses is that safety knowledge, safety attitude, and safety skills are all taught at the same time. The courses should be able to facilitate individual satisfaction, promote the understanding of technical knowledge but also of other people, foster healthy interpersonal relationships (on the same hierarchical level as well as on a different level), and ultimately lead to both more safety and higher productivity and production.

4.2 Process safety education (as an example)

4.2.1 Introduction

Following the increase in population and the accompanying worldwide rising demands of chemical products, chemical process installations are increasingly being built and exploited around the globe. Chemical industrial activities are always hazardous and accidents in such installations may lead to fatalities, environmental detriment, and huge economic losses. In brief, disasters may happen in the chemical industry. Similarly, a number of installations in the manufacturing industry and related to energy and fuel production store and/or process relevant quantities of dangerous chemical substances (e.g., gasoline, liquefied petroleum gas, molten salts used for cooling systems, etc.). Therefore, high-level competences are needed for those people operating and managing installations where relevant quantities of hazardous substances are present. The field of science dealing with the prevention, protection, and mitigation of consequences related to such installations and with the storage an manipulation of relevant quantities of hazardous substances is called "process safety."

The contribution of process safety education to the daily activities within the chemical process industry (CPI) is significant [5–11]. Process safety education is actually shaped and influenced by some major accidents in the CPI, in order to avoid the reoccurrence of these accidents. Apart from major accident prevention, process safety education also serves as the basis for process safety knowledge and know-how and the improvement of robust engineering practices in the CPI [12].

"Process safety education" refers to the learning of operating disciplines and safety culture through a systematic approach, with a view to preventing major accidents in the process industry. Process safety education is possible through three routes: (i) a university-based route, consisting of a bachelor's degree, a master's degree, and/or Ph.D. research (see Section 3.1); (ii) a professional route, consisting of internship, on-the-job training (OJT), continuous professional development (CPD), and/or industry-based research; and (iii) learning from inspections carried out by governmental regulatory agencies. This can be summarized as the "process safety education model," as illustrated in Figure 4.1.

In a university setting, process safety education begins with a bachelor's degree program or as a module that is embedded into undergraduate engineering programs such as chemical engineering [13, 14]. Process safety bachelor's programs introduce students to basic process safety principles and fundamental concepts and take between three and four years to complete. The subject can be studied further at a master's level, further developing the student skills and in-depth knowledge in a specialized area of process safety; this program is usually completed within two years. Lastly, for the interested and most promising students, a Ph.D.

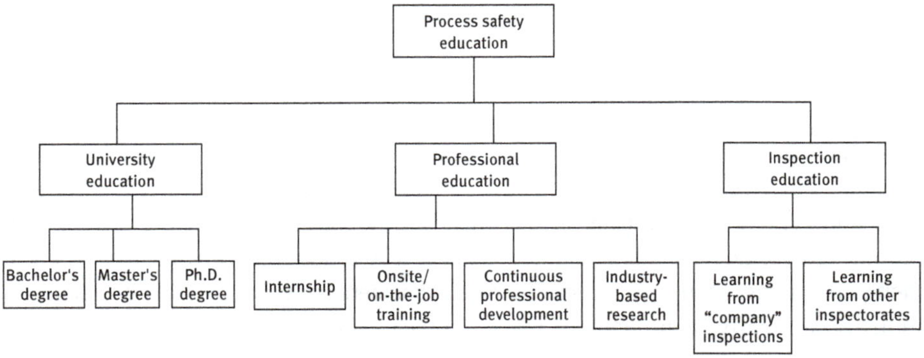

Figure 4.1: Process safety education model.

research program may be the final phase of a university-based process safety educational program. Such a program is fully focused on research, in the process safety domain, and is usually completed within three or four years. Further details on current models of process safety education are discussed in Chapter 5.

Professional training can be classified as the second phase of process safety education. It is performed within the industry and sometimes referred to as a "continuous learning" program. Process safety training may be categorized into four possible paths. First, internships can be organized enhancing students' exposure to industrial activities and further stimulating their theoretical knowledge. This is usually completed within three months to one year. Secondly, OJT, representing the learning from professional task execution, including initial training, retraining, and mentoring programs [15, 16], is a professional education path. A third path is called "continuous professional development" and is obtained from professional licensure, either as part of career capacity development or professional advancement; some examples are Professional Engineer (P.Eng.), Chartered Engineer (C.Eng.), Certified Functional Safety Expert (CFSE), and Professional Process Safety Engineer (PPSE) [17, 18]. The fourth path of process safety professional education is an industry-based research program, based on innovative experimental research and scientific observations within the industry. Such educational paths are strongly interrelated with academic education, as further discussed in Chapter 5.

The final route to process safety education is through inspection. Inspections can be company internal, external by consultants, or external by authorities. In Europe, for instance, by the inspections carried out under the obligations of the Seveso Directive (i.e., the inspections foreseen by the European Seveso Directive for the prevention of major accidents where hazardous substances are involved) are managed by the national competent authorities of each member state in the sites falling within the application of the Directive. Remark that inspectors are themselves subject to OJT.

Process safety education essentially constitutes lifelong learning in the field of safety designing, commissioning, operating, and decommissioning of installations of the chemical and process industry where relevant quantities of hazardous substances are present. Its aims to include, among others, an increased understanding of the process safety principles, advancement of technical proficiency, and promotion of knowledge sharing [19–21]. Apart from improving organizational performance, process safety education also promotes safety culture, increases productivity, sustains industry reliability, and enhances sustainable development in the process industry [22–27].

The main goal of process safety education is to provide to employees the competences for a safe and efficient operational conduct within the CPI. The term "competency" refers to professional capability, skills, and experience obtained through structured process safety education and training, in which employees' task performance is guided by an operational discipline. This concept encourages workers to perform tasks in a safe and efficient manner.

4.2.2 Process safety accreditation

Process safety education programs are often subject to an evaluation of their curricula by an accreditation body. Institutions offering process safety education are mandated to fulfil accreditation conditions laid out by the same body. Thus, engineering undergraduate programs and master's degree programs are endorsed by accreditation bodies, in line with set standards and guidelines. These conditions include checks on the curriculum, on the methodology, and on the learning outcomes.

The accreditation bodies designated for this role include the Accreditation Board for Engineering and Technology (ABET) in the United States of America, the Canadian Engineering Accreditation Board (CEAB) in Canada, the Institution of Chemical Engineers (IChemE) in the United Kingdom, the Engineers Australia Accreditation Board in Australia, as well as others.

The roles of these bodies can vary, but they often have the same goals, that concern, beside the accreditation of academic programs, in the update of requirements concerning competences and skills, in accordance with the evolving needs of society and of the companies interested in recruitment.

With respect to process safety core competences, a common attitude may be recognized for all the accreditation bodies when addressing chemical engineering or safety engineering programs. For example, ABET includes a discussion of the restructuring of process safety concepts in the curriculum of undergraduate chemical engineering programs [28]. Similarly, in Europe, plant and process safety education programs at both undergraduate and master's degree level are developed in accordance with the Bologna educational system [29, 30]. Furthermore, process safety education criteria are expected to adopt relevant programs from reputed process safety organizations [31]. The specific benchmarking activities of the IChemE Safety Center are expected to result in revisions of requirements concerning accreditation of chemical engineering programs by IChemE.

4.2.3 Process safety curricula

Curriculum development for process safety education has been proposed over the years by numerous authors. Such proposals include the integration of safety culture into the safety curriculum, consistency with accreditation (e.g., ABET) directives, adoption of recommendations, application of knowledge transfer, and so on [19, 32, 33]. The development of an adequate curriculum has also been cited as an essential requirement for effective process safety education [34]. Relevant contributions in curriculum development include the adoption of the Plant and Process Safety curriculum by Dechema (German Chemical Engineering Network) recommendation [35], and the development of a safety and loss prevention curriculum in chemical engineering programs of French universities [36]. Also, a special contribution has been made by Trevor Kletz on the curriculum development of inherently safer design of process installations [37].

Despite stakeholders' participation in developing process safety curricula, their elaboration and subsequent implementation within universities is often perceived as ineffective [38]. At present, generally accepted curricula for process safety education do not seem to exist in universities and industry [14, 39]. Currently, contents of process safety curricula applied in most universities include asset integrity and reliability, chemistry-related courses, fire and explosion studies, and process safety management. While a curriculum can be revised in the event of a process incident, as it was, for example, in the case of the so-called T2 laboratory reactive incident in the USA [40, 41], an "overcrowded" curriculum poses a threat to any revision [13, 42, 43]. Examples of process safety curricula will be further discussed in Chapter 5.

4.2.4 Teaching strategies for process safety

Teaching methodologies for process safety learning include storytelling, evidence-based teaching, and safety information sharing [39, 44, 45]. The purpose of the teaching strategies is to promote safety competency, technical design creativity, decision-making, and accident prevention [42]. As shown in Figure 4.2, and concurring with Figure 2.2, process safety education can be seen as an example of tier-based learning [46]. This approach can also be viewed in light of the process safety model of Figure 4.1: the (i) basic and (ii) applied principles correlate with (i) undergraduate

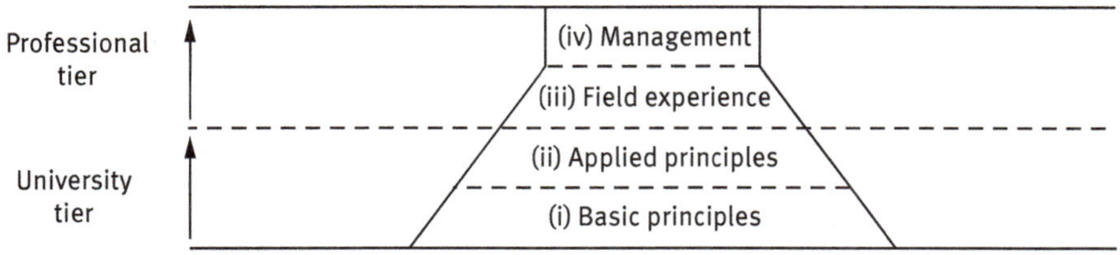

Figure 4.2: The learning process – learning pyramid model .
Source: Benintendi [46].

programs (eventually combined with internships) and (ii) a master's degree and Ph.D. program, respectively, while (iii) field experience and (iv) process safety management correlates with (iii) OJT and (iv) CPD, respectively.

However, recent studies show that there are only a limited number of universities where process safety is taught [14, 47]. This is obviously an impediment for the professional tier of Figure 4.2: if no foundation of basic and applied principles is laid at the university level, in order to guarantee adequate process safety knowledge and know-how, the principles need to be taught by the companies themselves before the professional tier can be commenced. The university tier of process safety is, however, sometimes not perceived by companies to be a core task which leads to recruit employees within such companies with no basic/applied competence, but only having learned about process safety from the viewpoint of the professional tier. The result is that process safety is often not adequately addressed in these organizations, since the learning process is partial when the basic and applied principles are not adequately understood by the learner.

4.2.5 Process safety research fields

Research in different fields related to process safety is a key tool in promoting the development of scientific knowledge and innovative technology in the process industry, so besides process safety education, adequate attention is required for process safety research and development. Research programs applicable to both university and industry comprise asset integrity and reliability, chemistry-related courses, design, hazard identification and risk analysis, human factors, as well as others (see Table 4.1). These research programs not only lead to technological or conceptual innovations, new theories, and models, but they also, for instance, promote competences and strengthen behavior-based safety [35, 48, 49]. Research centers are often established in universities to enhance scientific knowledge and promote industry's best practices.

Table 4.1: Process safety research.

Field of scientific research related to process safety	Reference
Fire and explosion studies, design	Gibson [48]
Chemistry-related courses, process control	Mannan et al. [79]
Safety performance and sustainability	Willey [53], Knegtering and Pasman [57], Pfeil et al. [51], Schonbucher et al. [30]
Risk analysis and management	Lundin and Jönsson [61]
Process safety education	Dadkhah [80], Mckay et al. [38], Schonbucher et al. [30]
Asset integrity and reliability, chemistry-related courses, design, hazard identification and risk analysis, human factor, incident management, risk management, safety culture, process safety management knowledge	Mannan [37], Amyotte [26], Pasman et al. [47]
Chemistry-related courses, design, fire and explosion studies, hazard identification and risk analysis, risk decision making, security, safety performance indicators	Mannan et al. [78]
Asset integrity and reliability, chemistry-related courses, design, fire and explosion studies, hazard identification and risk analysis, human factor, incident management, risk management, safety culture, safety management	Pasman et al. [47]
Safety management	Nesheim and Gressgård [21], Krause [14]
Hazard identification and risk analysis, incident management	Véchot et al. [81]
Design, hazard identification and risk analysis	Dee et al. [28], Meyer [49]

4.2.6 Process safety collaboration strategies

Process safety collaboration refers to cooperation among academia, industry, and authorities to foster the learning objectives required for the attainment of process safety. It should be noted that achieving effective collaboration requires substantial effort from university lecturers, industry professionals, accreditation bodies, governmental and regulatory agencies, and others. Moreover, partnerships in a wider scope can be developed by sharing information, internship opportunities, development of curricula, research opportunities, and funding (see also Table 4.2).

Aside from the enhancements gained by strategic involvement, other benefits of collaboration include the promotion of competences and the joint prevention of major accidents [50, 51]. Furthermore, the role of national and international

Table 4.2: Process safety education collaboration strategies mentioned in literature.

Collaboration strategies	Reference
Research	Gibson [48], Mannan et al. [79], Mannan [37], Mannan et al. [6], Lundin and Jönsson [61], Wu et al. [59]
CPI safety performance and sustainability	Osborn [81], Saleh and Pendley [42], Nesheim and Gressgård [21], Pasman et al. [47], Benintendi [46]
Funding	Mannan et al. [79], Schmidt [35]
Process safety publications	Pintar [82]
Curriculum development	CSB [40], Willey et al. [83], Mckay et al. [38], Crowl [33], Degreve and Berghmans [62], Amyotte [26], Spicer et al. [41], AIChE [56], Mannan et al. [84]
Internship	Dadkhah [80], Perrin and Laurent [36], Schmidt [35] Véchot et al. [81]
Shareholders (shares and assets)	Wasileski [72]
Teaching	Mannan et al. [79], Perrin and Laurent [36], Crowl [33], Degreve and Berghmans [62], Pitt [20], Saleh and Pendley [42], Schmidt [35], Véchot et al. [81], AIChE [56], Dixon et al. [43]

organizations in the development of a process safety curriculum are significant. Examples of such organizations are the American Institute of Chemical Engineers (AIChE), SAfety and Chemical Engineering Education (SAChE), ABET, the European Federation of Chemical Engineering (EFCE), and many more [41, 47, 52–55, 56].

Moreover, it is obvious that the interface among university, industry, and governmental agencies stimulates strategic collaboration objectives as funding, safety performance, curricula development, and research, among others. Figure 4.3 provides an overview of bilateral collaborations found in literature.

From Figure 4.3, it can be seen that cooperation between industry and government, on the one hand, and government and university, on the other hand, could be largely expanded. In an ideal situation, all the bilateral collaboration fields currently existing between university and industry in the scientific domain of safety, that is, curriculum development, funding, research, safety performance, publications, internships, and teaching, should also be present in the two other bilateral collaborationships. It should be remarked that several initiatives are presently addressing such issue, widening the scope in particular of collaboration among governmental organizations and academia. In several countries, specific curricula are offered for the preparation of inspectors and of other professional

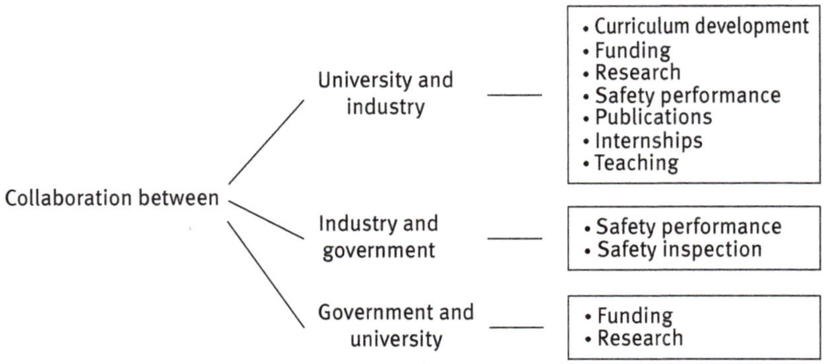

Figure 4.3: Process safety bilateral collaboration fields documented in the literature.

profiles addressing the needs of public authorities and public bodies, for example, firefighters or personnel of regional environmental agencies. Funding from public bodies is getting oriented more toward proposing the preparation of publications, dissemination material, or models and tools required by governmental agencies and control authorities. Internships in governmental organizations are frequently offered. Exchange of data for the preparation of lessons learnt is frequently offered by governmental agencies to academia. Thus, an evolving scenario is currently observed, going beyond the representation reported in Figure 4.3.

4.2.7 Process safety competence

Modern process installations are becoming more complex and their operation requires high-level competencies. Against this background, having a safety education curriculum that is combined with learning objectives is seen as necessary for efficient knowledge application [20, 23, 46, 57–60]. To meet the increasing need for process safety competencies, contributing factors such as research, funding, education, standards, and guidelines for best operating practices are very important [51]. These factors were recommended in a report published by the German Society for Chemical Engineering and Biotechnology/German Society for Process and Chemical Engineering (DECHEMA/GVC) (2004), in a recommendations report from the Dutch Hazardous Substances Council (2009), and also at the eighth edition of the European Congress of Chemical Engineering (ECCE) (2011). Furthermore, the promotion of process industry competencies have attracted the development of master's programs as witnessed at KU Leuven in Belgium and Lund University in Sweden [61, 62], as well as of several other initiatives, as discussed in detail in Chapter 5.

The use of appropriate learning methods is a requirement for improving process safety competencies [6]. Figure 4.4 illustrates the development of process safety

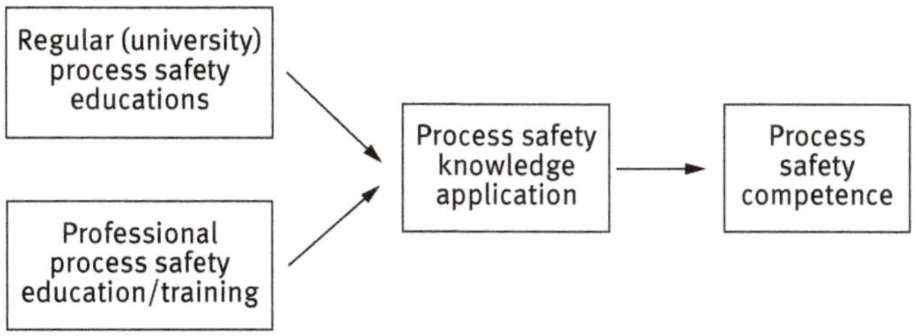

Figure 4.4: Achieving process safety competence.

competencies from knowledge application, gained through process safety education and training. This model conforms with a postulation on high-performance ranking resulting from education, experience, and training [21]. It is logical to assume that competence in task performance prevents major accidents in the chemical industry but more research is needed to confirm this assumption.

4.2.8 Process safety training

Training methods adopted by process safety training programs are intended to enhance the effectiveness of such programs. Apart from skill-based training, other methods include performance-based training, mentoring programs, team study, web-based training, and others [16, 63, 64]. These methods are regularly evaluated to assess training effectiveness [65, 66]. To meet the needs of the process industry, the update and restructuring of safety training is crucial [67]. Process safety training such as dynamic simulation, cognitive safety training, and hazard identification also helps the performance of operators in the process industry [68–74].

In the chemical process industry, process safety training requirements exist as a structured curriculum. The development of such a curriculum requires a concerted effort, most especially by the industry. A process safety training curriculum may include asset integrity and reliability, fire and explosion studies, hazard identification, risk analysis, and process safety management. Hereafter, an overview of process safety topics for professional training, mentioned in literature, is given. For an extensive overview, the interested reader is referred to Mkpat et al. [75]. Training curricula are developed according to industry needs [41] and are combined with learning aids to enhance training assimilation. The learning aids identified for effective training purposes in the process industries include multimedia, MP3 players, web-based teaching, computer-aided assessment, simulations, and more [25, 76]. The process safety topics mentioned in literature are asset integrity and reliability, chemistry related courses, design, economics, fire and explosion studies, hazard identification and risk analysis, human factor, incident management, process control, process safety management, regulation, risk decision making, security, and software programs.

4.2.9 Process safety education: a discussion

As it stands, the education of process safety is achieved by a combination of courses in universities and training in industry, and requires an unbiased analysis of its contributing elements. Although the accreditation bodies vary from one country to another, they ensure the standardization of process safety curricula used within universities. However, it has been observed that the accreditation of process safety education only applies to university-based programs, and does not exist in the industry and, more in general, for professional educational programs or training courses (professional tier in Figure 4.2). Thus, one may question whether all programs related to process safety education are adequately assessed and approved by accreditation bodies.

According to IChemE [60], accreditation bodies require universities to demonstrate the learning outcomes of their curriculum through the use of guidelines. This approach, first, demonstrates the credibility of such programs in sustaining the best industrial practices and, second, serves as a proof to the assessors within the accreditation process.

On the one hand, ABET in the USA demonstrates dominance in that country with most undergraduate engineering programs subscribing to it. ABET is now introducing requirements concerning process safety, in particular, within the accreditation of chemical engineering programs.

On the other hand, the Bologna Educational System is widely accepted in Europe, but with ineffective implementation across universities is missing an international accreditation body, and is not imposing any demands with respect to process safety.

Observation suggests that when curricula are utilized within university and industry, they have a similar content. It is also noted that curricula for both university and industry assign priority to hazard identification, risk analysis, incident management, and process safety management courses, so these courses can be ranked as highly important in gaining process safety knowledge. University curricula assign less priority to security and software programs, while professional curricula assign less priority to chemistry-related courses, economics, and less attention (similar to university curricula) to security, even when security threats are becoming prevalent [50]. It is against this backdrop that concepts such as design-based safety and collaboration in chemical industrial parks and clusters were developed [50, 77].

The teaching approaches of process safety help to strengthen knowledge application, improve safety culture, and foster prevention strategies. In most universities a strong preference for the integration of process safety courses into an existing program exists, as opposed to teaching process safety as a dedicated course. This situation is attributed to a lack of trained professional lecturers in this field of study [20]. To address this gap, professionals from industry could support universities by teaching process safety as a dedicated course [43], as is the case for instance in the process safety program developed at the KU Leuven.

Learning aids have become increasingly relevant in the advancement of process safety education both in university and in industry. These learning aids promote knowledge assimilation and improve effective knowledge transfer in the process industry. They also exist as process safety–related products, process safety–related software programs, and process safety regulations. The initiative of developing process safety products has been strongly supported by international organizations. For example, as already mentioned, in the USA there exist the Center for Chemical Process Safety, SAChE, the US Chemical Safety and Hazard Investigation Board (CSB), the Mary Kay O' Connor Process Safety Center, and AIChE. Similarly, in Europe, the following organizations can be found: the European Process Safety Centre, EFCE, IChemE, and the Safety to Safety (S2S) initiative. In Canada the following institutions are present: the Canadian Society for Chemical Engineering and Minerva Canada Safety Management Education.

In the process industry, the performance of tasks is expected to be as efficient as possible, and this is dependent on process safety competencies. Process safety competence can be understood as the main objective of the process safety education model. In this model, competence is identified as a key element, and is defined at all levels from the bachelor's degree in university education to that of an inspector from a governmental agency.

Universities tend to adopt process safety curricula different from one another, despite offering the same program [14, 39]. This situation also applies to process safety professional training in the industry [41]. Hence, tailor-made curricula, rather than one generic curriculum, are found in academic and industrial practice. It is logical to conclude that tailor-made curricula address specific needs and are not holistic; hence, competence demonstration is constrained. To meet this need, further research needs to be conducted in order to verify whether a harmonized curriculum might not be a better option than the tailor-made curricula of today.

Furthermore, accreditation of the process safety curriculum in most universities lacks effective implementation, and adjusting existing programs to accommodate process safety accredited courses are met with resistance because of existing, overloaded curricula [13, 42, 43]. A further recommendation is the industry identifying its proper needs and collaborating with accreditation bodies and universities for effective development and implementation of the process safety curriculum.

4.3 Conclusion

The experience of process safety education is far from peculiar. In several other industrial sectors besides the chemical industry, high safety standards are required:

for example, the nuclear industry, the aeronautical industry, the construction industry when dealing with infrastructures or large residential structures, and so on.

In all industrial sectors, trained professionals with high regard for safety culture and specific safety competence are needed by the industry and by control authorities.

The paradigm of process safety can be applied to most of such sectors. The patterns concerning educational models shown in Figure 4.1, education tiers represented in Figure 4.2, bilateral collaborations represented in Figure 4.3, and the process toward achieving safety competences shown in Figure 4.4 apply as well to all other industrial sectors and engineering disciplines where a high safety competence is required.

The problems concerning the harmonization of the contents of the courses, the question of teaching safety as an independent course or embedded within other subjects, and the issue of accreditation apply worldwide to all other engineering disciplines.

This should not be surprising. Actually, safety is an interdisciplinary subject and, in all industrial sectors, safety concerns came after the boost of technology development that took during the industrial revolution, between 1850 and 1950. It was only after the Second World War that safety concerns evolved from the protection of workers to the protection of the society from industrial hazards.

Thus, it should not be surprising that in most educational programs, set up in the tradition of educating engineers to understand and develop technology, teaching safety was initially an "add-on" within an established program. Even if several decades passed since this situation, safety education models are still jeopardized and grew up mostly in a local context, with no international benchmarking or harmonization. The need for a more standardized educational profile of ERManagers is, however, now perceived by all stakeholders. This may help to step forward in identifying core contents and shared educational models, starting from the wide experiences and the existing educational programs, as those discussed in the following chapter.

References

[1] Meyer, T., Reniers, G. (2016). *Engineering Risk Management*. De Gruyter: Berlin.

[2] Reniers, G., Van Erp, N. (2016). *Operational Safety Economics*. Wiley-Blackwell: London.

[3] Petersen D. (1978), Safety Management. A human approach. Aloray Inc.: New York.

[4] Setty E.D., Moorthy P.K. (2008). Training methods focusing on industrial safety. Akansha Publishing House: New Delhi.

[5] Hurme, M., Rahman, M. (2005). Implementing inherent safety throughout process lifecycle. *Journal of Loss Prevention in the Process Industries*, 18 (4–6), 238–244.

[6] Mannan, M.S., Qi, R., Prem, K.P., Ng, D., Rana, M.A., Yun, G. (2012). Challenges and needs for process safety in the new millennium. *Process Safety and Environmental Protection*, 90 (2), 91–100.

[7] Aziz, H.A., Shariff, A. M., Rusli, R., Yew, K. H. (2014). Managing process chemicals, technology and equipment information for pilot plant based on Process Safety Management standard. *Process Safety and Environmental Protection*, 92(5), 423–429.
[8] Hopkins, A. (2015). The cost–benefit hurdle for safety case regulation. *Safety Science*, 77, 95–101.
[9] Schenk, L., Antonsson, A.-B. (2015). Implementation of the chemicals regulation REACH – Exploring the impact on occupational health and safety management among Swedish downstream users. *Safety Science*, 80, 233–242.
[10] Majid, N.D., Mohd Shariff, A., Mohamed Loqman, S. (2016). Ensuring emergency planning & response meet the minimum Process Safety Management (PSM) standards requirements. *Journal of Loss Prevention in the Process Industries*, 40, 248–258.
[11] Swuste, P., Reniers, G. (2016). Seveso inspections in the European low countries history, implementation, and effectiveness of the European Seveso directives in Belgium and the Netherlands. *Journal of Loss Prevention in the Process Industries*, 49 (A), 68–77.
[12] Guntzburger, Y., Pauchant, T.C., Tanguy, P.A. (2016). Ethical Risk Management Education in Engineering: A Systematic Review. *Science and Engineering Ethics*, 23(2), 323–350.
[13] Hendershot, D.C., Louvar, J.F., Kubias, F.O. (1999). Add chemical process safety to the chemistry curriculum. *Chemical Health and Safety*, 6 (1), 16–22.
[14] Krause, U. (2016). Process Safety in Engineering Education – Pro's and Con's of Different pproaches. *Chemical Engineering Transactions*, 48, 871–876.
[15] Crowl, D., Louvar, J. (2002). *Chemical Process Safety, Fundamentals with Applications*, Prentice Hall: New Jersey.
[16] Young, C. W., Hodges, K. J. (2012). Process safety management mentoring: Subjects to convey and the methods for conveying. *Process Safety Progress*, 31 (4), 350–354.
[17] Exida. (2017). Exida Training – Functional Safety, ICS Cybersecurity, Alarm Management, CFSE. Retrieved April 3, 2017, from http://www.exida.com/Training.
[18] IChemE. (2017). IChemE Safety Centre | Training. Retrieved April 2, 2017, from http://www.ichemesafetycentre.org/isc-training.aspx
[19] Louvar, J., Hendershot, D. (2007). Education materials for universities and industry. *Process Safety Progress*, 26 (2), 85–89.
[20] Pitt, M.J. (2012). Teaching Safety in Chemical Engineering: What, How and Who? *Chemical Engineering & Technology*, 35 (8), 1341–1345.
[21] Nesheim, T., Gressgård, L.J. (2014). Knowledge sharing in a complex organization: Antecedents and safety effects. *Safety Science*, 62, 28–36.
[22] Cox, S.J., Tart, N.R. (1991). *Reliability, Safety and Risk Management; An integrated Approach*. Butterworth-Heinemann: Oxford.
[23] Elangovan, R.K., Mohammed, K.P., Mohan, S. (2005). Effectiveness of the designed safety education programme modules by their implementation in selected industries. *Journal of Loss Prevention in the Process Industries*, 18 (4–6), 553–557.
[24] Sutton, I.S. (2008). Use root cause analysis to understand and improve process safety culture. *Process Safety Progress*, 27 (4), 274–279.
[25] Myers, P.M., Watson, B., Watson, M. (2008). Effective training programs using instructional systems design and e-learning. *Process Safety Progress*, 27 (2), 131–138.
[26] Amyotte, P.R. (2013). Process safety educational determinants. *Process Safety Progress*, 32 (2), 126–130.
[27] Kluge, A., Nazir, S., Manca, D. (2014). Advanced Applications in Process Control and Training Needs of Field and Control Room Operators. *IIE Transactions on Occupational Ergonomics and Human Factors*, 2 (3–4), 121–136.

[28] Dee, S.J., Cox, B.L., Ogle, R.A. (2015). Process safety in the classroom: The current state of chemical engineering programs at US universities. *Process Safety Progress*, 34 (4), 316–319.

[29] Cortés, J.M., Pellicer, E., Catala, J. (2011). Integration of occupational risk prevention courses in engineering degrees: Delphi study. *Journal of Professional Issues in Engineering Education & Practice*, 138 (1), 31–36.

[30] Schonbucher, A., Brenig, H.W., Klais, O., U. Hauptmanns, J. Schmidt, Moritz, H.U. (2013). *Model and Curriculum "Process and Plant Safety."* Frankfurt, M: Dechema.

[31] Amyotte, P.R., Berger, S., Edwards, D.W., Gupta, J.P., Hendershot, D.C., Khan, F.I., ... Willey, R.J. (2016). Why major accidents are still occurring. *Current Opinion in Chemical Engineering*, 14, 1–8.

[32] Hendershot, D.C., Smades, W. (2007). Safety culture begins in the classroom. *Process Safety Progress*, 26 (2), 83–84.

[33] Crowl, D. (2012). Process Safety Education: Meeting the New ABET Requirements. *Chemical Engineering Progress*, April, 19.

[34] Ferjencik, M. (2007). Best starting point to comprehensive process safety education. *Process Safety Progress*, 26 (3), 195–202.

[35] Schmidt, J. (2013). Process and Plant Safety – Research & Education Strategy to Keep Long Term Competences. *AIDIC*, 13, 421–426.

[36] Perrin, L., Laurent, A. (2008). Current situation and future implementation of safety curricula for chemical engineering education in France. *Education for Chemical Engineers*, 3 (2), e84–e91.

[37] Mannan, M.S. (2012). Trevor Kletz's impact on process safety and a plea for good science – An academic and research perspective. *Process Safety and Environmental Protection*, 90 (5), 343–348.

[38] McKay, G., Noakes, N., Chow, C.C.L., Ko, E. (2011). Safety education for chemical engineering students in Hong Kong: Development of HAZOP Study teaching module. *Education for Chemical Engineers*, 6 (2), e31–e55.

[39] Rae, A. (2016). Tales of disaster: the role of accident storytelling in safety teaching. *Cognition, Technology & Work*, 18 (1), 1–10.

[40] CSB. (2009). U.S. chemical safety and hazard investigation board investigation report t2 laboratories, inc. Runaway reaction. U.S Chemical Safety and Hazard Investigation Board (CSB). Retrieved from http://www.csb.gov/t2-laboratories-inc-reactive-chemical-explosion/

[41] Spicer, T.O., Willey, R.J., Crowl, D.A., Smades, W. (2013). The safety and chemical engineering education committee-broadening the reach of chemical engineering process safety education. *Process Safety Progress*, 32 (2), 113–118.

[42] Saleh, J.H., Pendley, C.C. (2012). From learning from accidents to teaching about accident causation and prevention: Multidisciplinary education and safety literacy for all engineering students. *Reliability Engineering & System Safety*, 99, 105–113.

[43] Dixon, D.J., Kohlbrand, H.T. (2015). Lending industrial experience through reactive hazard examples in university safety instruction. *Process Safety Progress*, 34 (4), 360–367.

[44] Shallcross, D.C. (2013). Safety education through case study presentations. *Education for Chemical Engineers*, 8 (1), e12–e30.

[45] Cheah, S.-M. (2016). Evidence-Based Flipped Classroom Case Study – Teaching Chemical Process Safety. *International Symposium on Advances in Technology Education*. Retrieved from https://www.researchgate.net/publication/308265915_EVIDENCE-BASED_FLIPPED_CLASSROOM_CASE_STUDY_-_TEACHING_CHEMICAL_PROCESS_SAFETY

[46] Benintendi, R. (2016). The bridge link between university and industry: A key factor for achieving high performance in process safety. *Education for Chemical Engineers*, 15, 23–32.

[47] Pasman, H.J., De Rademaeker, E., Suter, G., Fabiano, B. (2014). A review of the past, present and future of the European loss prevention and safety promotion in the process industries. *Process Safety and Environmental Protection*, 92 (4), 280–291.
[48] Gibson, N. (1999). Process safety–a subject for scientific research. *Process Safety and Environmental Protection*, 77 (3), 149–153.
[49] Meyer, T. (2015). Towards the implementation of a safety education program in a teaching and research institution. *Education for Chemical Engineers*, 18, 2–10.
[50] Reniers, G., Amyotte, P. (2012). Prevention in the chemical and process industries: Future directions. *Journal of Loss Prevention in the Process Industries*, 25 (1), 227–231.
[51] Pfeil, N., Jochum, C., Mitropetros, K., Schmelzer. (2013). Keeping and Improving Process and Plant Safety Competence – What is Needed, What Should be Done? *Chemical Engineering Transactions*, 31, 373–378.
[52] Bryan Jr, L.A. (1999). Educating engineers on safety. *Journal of Management in Engineering*, 15 (2), 30–33.
[53] Willey, R.J. (1999). SACHE Case Histories and Training Modwles. *Process Safety Progress*, 18 (4), 195–200.
[54] Louvar, J.F., Hendershot, D.C. (2003). SACHE: 17 years of promoting teaching of safety to chemical engineering students. *Chemical Health and Safety*, 10 (5), 8–10.
[55] Louvar, J.F. (2009). Safety and chemical engineering education-History and results. *Process Safety Progress*, 28 (2), 131–134.
[56] AICHE. (2015). Dow makes leadership gift to launch major process safety education initiative. *Chemical Engineering Progress*, December, 53.
[57] Knegtering, B., Pasman, H.J. (2009). Safety of the process industries in the 21st century: A changing need of process safety management for a changing industry. *Journal of Loss Prevention in the Process Industries*, 22 (2), 162–168.
[58] EFCE. (2010). Recommendations for Chemical Engineering Education in a Bologna Three Cycle Degree System. EFCE, August.
[59] Wu, T.-C., Chang, S.-H., & Chen, D.-F. (2012). Developing a competency model for safety professionals: Correlations between competency and safety functions. *Journal of Safety Research*, 43 (5–6), 339–350.
[60] IChemE. (2015). Accreditation of Chemical Engineering Programmes – A guide for higher education providers and assessors. IChemE, November.
[61] Lundin, J., & Jönsson, R. (2002). Master of science in risk management and safety engineering, at Lund University, Sweden. *Journal of Loss Prevention in the Process Industries*, 15 (2), 111–117.
[62] Degreve, J., Berghmans, J. (2012). Master of Science in Safety Engineering at KU Leuven, Belgium. *Procedia Engineering*, 45, 276–280.
[63] Cusimano, J. (1995). Chemical Plant Empowers Work Force To Initiate Process Safety Training. *Occupational Health and Safety*, 46.
[64] Taylor, M.A., Wirth, O., Olvina, M., Alvero, A.M. (2016). Experimental analysis of using examples and non-examples in safety training. *Journal of Safety Research*, 59, 97–104.
[65] Cooper, M., Cotton, D. (2000). Safety training – a special case? *Journal of European Industrial Training*, 24(9), 481–490.
[66] Haesle, J., Devlin, C., Mccavit, J.L. (2009). Improving process safety by addressing the human element. *Process Safety Progress*, 28 (4), 325–330.
[67] Louvar, J.F. (2008). Improving the effectiveness of process safety management in small companies. *Process Safety Progress*, 27 (4), 280–283.

[68] Hub, L. (1999). Simulation Program for Hazard Training and Safety Evaluation of Chemical Processes. *Chemical Engineering Technology*, 22 (9), 740–742.

[69] Gatfield, D. (1999). Can cognitive science improve the training of industrial process operators? *Journal of Safety Research*, 30 (2), 133–142.

[70] Cann, N. (2001). Use of the safety case as a training tool to disperse corporate knowledge. *IChemE*, 148, 655–663.

[71] Shacham, M., Eizenberg, S., Brauner, N. (2006). Combining HAZOP with dynamic simulation – Applications for safety education. *Journal of Loss Prevention in the Process Industries*, 19 (6), 754–761.

[72] Wasileski, R.F. (2009). Spot the Hazard! A cultural extension of hazard identification training. *Process Safety Progress*, 28 (2), 200–206.

[73] Nazir, S., Manca, D. (2015). How a plant simulator can improve industrial safety. *Process Safety Progress*, 34 (3), 237–243.

[74] Yamamoto, S. (2015). A systematic knowledge education approach for safety-critical system development. *Procedia Computer Science*, 60, 960–967.

[75] Mkpat, E., Reniers, G., Cozzani, V. (2018). Process safety education: A literature overview. *Journal of Loss Prevention in the Process Industries* 54, 18–27.

[76] Moon, I., Goh, S., Chang, B., Jeong, I., Kwon, H.-T. (1998). Safety improvement by a multimedia operator education system. *Computers and Chemical Engineering*, 22, S531–S536.

[77] Reniers, G., Khakzad, N. (2017). Revolutionizing safety and security in the chemical and process industry: Applying the CHESS concept. *Journal of Integrated Security Science*, 1, 2–15.

[78] Mannan, S., Chowdhury, A., Reyes-Valdes, O. (2012). A Portrait of Process Safety: From Its Start to Present Day. *Hydrocarbon Processing*, 91 (7), 55–62.

[79] Mannan, S., Akgerman, A., Anthony, R., Darby, R., Eubank, P., Hall, K. (1999). Integrating process safety into chE education and research. *Chemical Engineering Education*, 33 (3), 198–209.

[81] Osborn, L. (1999). Process Safety in Education. *Process Safety Progress*, 18 (4), W5.

[80] Dadkhah, A.A. (2008). A Study on the Inclusion of the Process Safety in Chemical Engineering Curriculum and Student responses at Isfahan University of Technology. Retrieved from http://dadkhah.iut.ac.ir/sites/dadkhah.iut.ac.ir/files/file_pubwdet/2008-dadkhah-iransafety-isfahan-10200179.pdf

[81] Véchot, L.N., Casson, V., Olewski, T. (2014). Training future engineers to be committed to safety. *QScience Proceedings*, 2014 (3), 28.

[82] Pintar, A.J. (1999). Teaching Chemical Process Safety: A Separate Course versus Integration into Existing Courses. *Age*, 4, 1.

[83] Willey, R.J., Fogler, H.S., Cutlip, M.B. (2010). The integration of process safety into a chemical reaction engineering course: Kinetic modeling of the T2 incident. *Process Safety Progress*, n/a-n/a.

[84] Mannan, M.S., Chen, H., Pittman, W.C., Hatanaka, L.C., Harding, B.Z., Boussouf, A., … Milke, J.A. (2015). Integration of process safety engineering and fire protection engineering for better safety performance. *Journal of Loss Prevention in the Process Industries*, 37, 74–81.

5 Example of safety or ERM education programs, courses and contents

5.1 University programs

5.1.1 Skills required in the educational background of ERManagers

In Chapter 3, the competences of an engineering risk manager (ERManager) were discussed in detail. Four main categories, recalled below, may be used to summarize them:
1. *Technical skills*, sufficient to understand the managed systems, the hazards and risks deriving from such systems, and the safety barriers available for risk management
2. *Economic/managerial skills*, sufficient to carry out, for example, cost–benefit analyses and other assessments required for decision-making
3. *Social skills*, sufficient to interact with other technical disciplines, to plan safety education and training, and to promote safety commitment and a positive safety climate
4. *Specific skills*, related to the understanding of state-of-the-art structured techniques for hazard and risk assessment, risk prevention and mitigation, risk monitoring and control

Such competences are usually obtained by an articulated learning process, which in general starts in academia, as mentioned in Chapters 3 and 4. It is important to have in mind that all these four categories of skills need to be provided to an ERManager during the education process.

5.1.2 Competence providers

Two main categories of players are usually involved in the education process of an ERManager as competence providers: universities and specialized companies. The latter, however, in general provide only professional courses that may contribute to the acquisition of the specific skills mentioned above, thus having a relevant role only in the final part of the learning process (the professional tier in the learning pyramid represented in figure 4.2).

It is also important to remark that universities also have an important role in providing professional courses, directly or participating in companies or consortia

https://doi.org/10.1515/9783110344578-005

providing educational services. At the same time, academics may be involved in professional courses provided by specialized companies.

Beside these two main players that usually organize, offer, and promote the programs that support the learning process of the ERM, a third important category of players is generally involved in the learning process, in particular, when specific and focused skills need to be provided, usually in the final part of the formation. These are often indicated as “knowledge owners”: professionals having a huge experience on ERM due to their present or past positions as an ERManager in industrial or consultancy companies.

Knowledge owners may be directly involved in teaching. When available, they are usually called both by academia and by specialized companies to deliver classes or to take the responsibility of specific teachings within academic or professional programs. Besides conventional teaching methods, the on-field first-hand experience of “knowledge owners” allows a more effective application of interactive teaching methods, as not only the analysis of case histories and of lessons learnt, but also specific role playing experiences.

The direct involvement of knowledge owners in delivering academic or professional programs in ERM is, however, far from simple. Knowledge owners need to be involved in the design of the course and to rationalize their experience in preparing effective case histories and/or simulation scenarios. An effective preparation of teaching material and of classes usually requires the interaction with teaching professionals or academics, in order to check and validate the design of the learning cases, and to warrant a realistic timing in the delivery suitable for an effective learning process. Moreover, obtaining specific authorizations may be necessary to knowledge owners in order to be allowed disclosing case histories, lessons learnt, or even the technical details needed to describe the technical system of concern.

A more simple access to a first-hand real-life professional learning experience and to a mentorship by knowledge owners may be obtained by internship or training-on-the-job experiences. Mentoring by knowledge owners applied to real-life problem solving is actually a very efficient learning experience (“learning by doing”) and is easier to afford by knowledge owners, requiring less preparatory work. Furthermore, the company hosting the learning experience may benefit from the transfer of “fresh” ideas and independent reasoning models from the trainee. However, a limitation of such experiences is that the background of the learner should be sufficient to afford the real-life problems usually tackled within such experiences. Time available for mentorship is usually limited and higher skills, higher independence, and a more active contribution from the learned is usually required than in classes, in role playing, and/or in the guided analysis of case studies. In general, higher level academic programs usually include or require such experiences, in particular, as a last step of the learning process within the university tier of the learning pyramid represented in Figure 4.2.

5.1.3 General framework of university programs for ERM education

In order to understand the different academic programs available in the framework of ERM, it is important to define a detailed framework of the educational processes available for the formation of an ERManager. As discussed in Chapter 3, the educational process ideally starts in academia. Figure 5.1 provides not only a realistic but also an idealized scheme of the two more frequent tracks in the education of an ER-Manager. As shown in the figure, in both tracks the educational process starts in academia, attending a bachelor's in engineering disciplines. Depending on the geographical location and on the university, the bachelor's (step 1 in Figure 5.1) may last 3 or 4 years. The specific engineering discipline (e.g., aerospace engineering, chemical engineering, structural engineering, etc.) chosen is usually the more pertinent to the industrial activities were the ERManager is likely to operate, in perspective, at least in the early steps of his/her professional career.

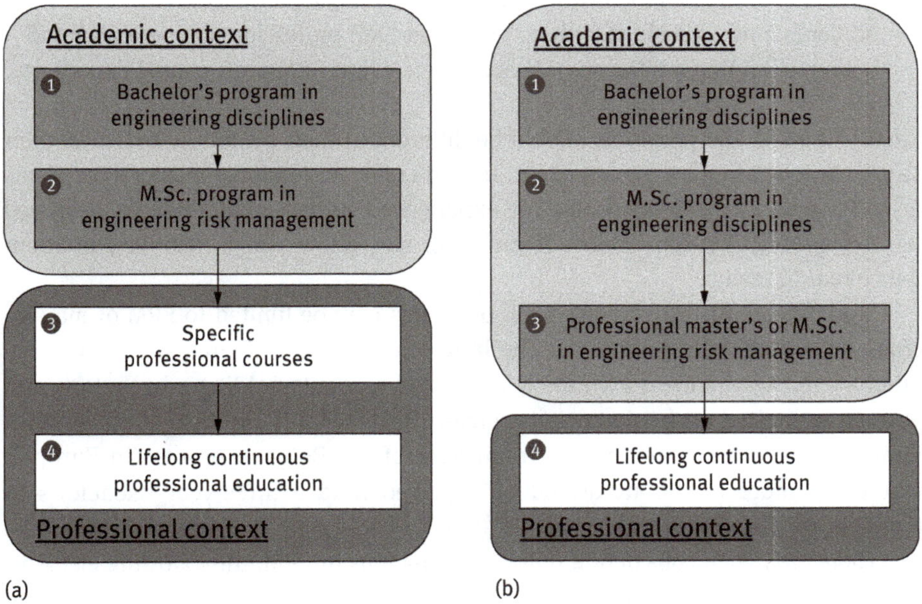

Figure 5.1: Scheme of the two more frequent tracks ideally leading to the formation of an ERManager. Track (a): Education on risk management starts attending a specific M.Sc. program in an academic framework after completing a bachelor's in engineering disciplines; Track (b): Education on risk management starts after completing a bachelor's and a master's program in engineering disciplines.

After this first step, Figure 5.1 shows two alternative possibilities: in track (a), an academic 1 or 2 years M.Sc. program addressing the risk management framework is attended (step 2). The educational profile is then completed attending specific courses (steps 3 and 4), usually in a professional context. As an alternative (track (b) in Figure 5.1), following the bachelor's, the educational process continues within engineering disciplines. Competence on the risk management framework is obtained only after completing step 2, attending a specific professional master's or M.Sc. program (step 3 of track b). This is more usually attended at academia rather than in a professional context, although some companies propose structured professional master's courses on ERM to their employees.

When considering both tracks in Figure 5.1, it should be considered that it is quite common that the learners will start step 2 immediately after ending step 1, while it is rather frequent that step 3 is undertaken only after a few years of professional experience. This applies, in particular, to professional master's programs, which are usually attended only after acquiring a few years of professional experience.

Therefore, when proposing M.Sc. programs or professional master's programs addressing ERM, universities should design the course in order to respond to one or more of the following potential attendees:

- students holding a B.Sc. with no ERM professional experience;
- students holding a M.Sc. with no or limited ERM professional experience; and
- students holding a B.Sc. or an M.Sc. with relevant ERM professional experience.

Clearly enough, the programs should be different in their content and also in their structure depending on the attendees. In particular, professional master's programs or M.Sc. programs intended also for experienced professionals may be designed (also or specifically) for distance learning, or may allow a high flexibility in attendance requirements.

Moreover, skills provided in such programs may be limited to ERM or may include as well technical contents. The latter are usually delivered within the program when the attendees are mostly graduated students holding a bachelor's and with no or limited professional experience. This is usually the case of universities delivering M.Sc. programs in the framework of the Bologna process(in Europe), where the more numerous attendees expected hold a three-year bachelor's in engineering disciplines.

Therefore, when considering university programs in ERM, the following elements should be considered in order to correctly understand and compare the educational project:

- *Type of program*: M.Sc. programs or professional programs
- *Duration*: 12 or 24 months
- *Admission requirements*: generic engineering background, specific engineering background, bachelor's or M.Sc. level, and so on
- *Primary attendees*: students holding a bachelor's, students holding a master's, professionals
- *Attendance*: part time, full time, and so on

- *Learning mode*: distance learning, blended
- *Type of contents*: technical, economic, social, specific (as discussed above)

5.2 Different examples from different countries

It is evident from the above discussion that programs dedicated to the education of ERManagers may be very different, even when limiting the analysis to those run in the academic context. It is a challenging task to frame a comprehensive overview of the university programs dedicated to the education of ERManagers.

Therefore, in the following section, only a few representative examples from different geographical areas are reported.

It should be remarked that the examples were selected only for demonstration purposes, in order to give evidence of the different educational tracks present and to discuss their contents and the differences present among different types of programs. It should thus be considered that similar programs beside those discussed may be offered by other universities. It should also be considered that programs are continuously updated. Therefore, even if an effort was made to include in the discussion updated and representative information, the reader is referred to the websites of the universities to obtain updated information on program structures and contents.

In Europe a number of rather different programs are proposed for the education of ERManagers. In continental Europe and in the Scandinavian regions, most educational programs are offered as second-cycle degrees in the framework of the Bologna process. Thus, the programs are mostly two-year full-time masters, in general requiring a bachelor's in engineering disciplines for admission. In UK or US universities, 18-month programs are the prevailing model: two terms dedicated to courses and the third term to carry out a research project in the university or during an internship in a company.

The balance among the different disciplines may be rather different in different initiatives, as well as the focus of the educational process. In the following section, examples are given addressing the following specialization areas:
- industrial risk management;
- reliability and safety management;
- process safety management;
- disaster risk management; and
- environmental risk management.

Even if the examples provided range on several specialization areas, due to editorial limitations much more are left out: for example, in specialization areas as nuclear industry, aeronautical industry, fire protection, and so on, important and well-reputed educational programs addressing risk management are as well available. However, the purpose of this chapter is only to exemplify how competence on ERM is built. The examples provided have the purpose to identify and exemplify the more common educational patterns that are applied to build a professional profile of an ERManager, irrespective of the area of specialization addressed. Thus, the patterns identified are mostly applicable also to other disciplinary areas where ERManager specialists are required.

5.2.1 Industrial risk management

Programs focusing specifically on the formation of "general" risk managers, that is, not focused on specific technical areas, are offered by a few universities. For example, Table 5.1 lists the outline of the programs running at the University of Stavanger in Norway [1]. As listed in the table, in the first year four fundamental groups of disciplines are addressed in the formation: fundamentals of mathematics for risk management (mainly probability and statistics), fundamentals of reliability, fundamentals of economics, and introduction to risk analysis and risk management. In the second year, education to specific risk management in alternative disciplinary specialization areas is addressed to introduce the specific technical elements and risk management tools needed in the selected specialization area.

Table 5.1: Scheme of the two-year master's program on risk assessment and management offered in AY 2018–2019 by the University of Stavanger (Norway).

Risk assessment and management **Two-year master's, full time, in presence**
First year
Risk analysis and risk management Introduction to probability and statistics Reliability analysis Economic analysis in risk management
Second year: alternative specialization areas
Cyber Security Offshore Technical Safety Risk Assessment and Management Reliability and Resilience Engineering
Master's thesis to be selected within the specific specialization area selected

A similar approach is proposed by the University of Aalborg (Denmark), as shown in Table 5.2 [2]. Also in this case the required courses are in four main areas: fundamentals of mathematics for risk management (mainly probability and statistics),

Table 5.2: Scheme of the two-year master's program on risk and safety management offered in AY 2018–2019 by the University of Aarlborg (Denmark).

Risk and safety management **Two-year master's, full time, in presence**
First year
Systems engineering Applied statistics and probability theory Risk analysis Industry standards and legislation (project) Risk management Decision making Choose between risk communication and maintenance management Risk analysis and management (project)
Second year
Simulation of emergencies Emergency management Choose between risk and reliability in engineering and health and safety management Operational risk management in projects (project)
Master's thesis (30 ECTS) to be selected within a specialization area

fundamentals of reliability (and of maintenance in the specific case), introduction to risk analysis and risk management, and operational risk management. The fourth area is slightly wider than that in Stavanger, since operational risk management and decision-making include, but are not limited to, fundamentals of economics. However, it should be remarked that in this program the general formation provided is wider, since required courses are present also in the first semester of the second year and specialization on a specific disciplinary area is limited to the thesis project carried out in the second semester of the second year.

A one-year full-time master's in Industrial Risk Management is offered by MINES – ParisTech [3]. As shown in Table 5.3, also in this initiative, the program starts with a theoretical part where general risk management concepts are delivered and prosecuted with a specialization addressing specific sectorial needs. Since the duration is limited to one year, the contents of the theoretical program are more focused, addressing mostly methods and tools for crisis management and operational risk management, thus joining social theory and practice, engineering, legal, and management sciences. The part addressing sectorial needs is conceived as an extended internship in a company that is concluded by the preparation of a final thesis. The initiative is certified by the French Conference des Grandes Ecoles.

In the frame of general educational programs for risk managers, a very specific initiative is run by the Risk Research Center at ETH Zurich (Switzerland). The

Table 5.3: Scheme of the master's program in industrial risk management offered in AY 2018–2019 by MINES-Paritech (France).

Industrial risk management **One-year master's, full-time, in presence**
First period (6 months) – theoretical program
Methods and tools for crises management Organization behaviors Management of different situation types Team work on field Industrial visits
Second period (6 months) – professional mission
The professional mission is a specific study performed in a company. The student will be supervised by a professor and a representative from the company. The mission will end with a professional thesis and an oral defense.

Center offers a certification on "MSc studies with a focus on Risk Management and Policy" [4]. The certification may be obtained completing any master's degree in engineering disciplines at ETH, provided that the four conditions listed in Table 5.4 are fulfilled. The idea is to prepare students for a better management of risk and enabling them to design novel solutions for risk management directly during their education in engineering disciplines at a master level. As evident

Table 5.4: Scheme of the requirements to be fulfilled to obtain the Risk Studies Certificate "MSc studies with a focus on Risk Management and Policy" in AY 2018–2019 by the Risk Center and the Department of Management, Technology and Economics at ETH Zurich (Switzerland).

Risk Studies Certificate "MSc studies with a focus on Risk Management and Policy" obtained by fulfilling the below requirements while completing a two-year full-time in-presence master's in engineering disciplines
1. A tutor should be selected among the members of the ETH Risk Center (chairs are currenty on the following topics: integrative risk management; macroeconomics: innovation and policy; risk and insurance economics; system design; and entrepreneurial risks). 2. A total of 24 ECTS should be obtained by selecting and completing a set of courses addressing topics comprised within the following disciplinary areas: Information Security; International Conflict Research; Mathematics; Land Use Engineering; Computational Physics; Reliability and Risk Engineering; Structural Dynamics and Earthquake Engineering; Risk, Safety and Uncertainty Quantification; and Traffic Planning and Transport Systems. 3. An internship (6 ECTS) in the area of risk management should be obtained. 4. The master's thesis should address a risk-related topic.

from Table 5.4, in this case the education is mostly oriented to provide knowledge concerning the specific elements of risk management in the specialization area selected by the student.

5.2.2 Reliability and safety management

Several universities offer educational programs addressing ERM in the framework of reliability and maintenance. For example, NTNU Trondheim offers a two-year M. Sc. Program in Reliability, Availability, Maintainability and Safety (RAMS) based on the scheme listed in Table 5.5 [5]. As shown in the table, the program is more focused on engineering disciplines, including maintainability and systems engineering as required courses, and does not include safety economics or statistics as separated required courses (even if elements of both topics are embedded in the required courses). Communication and self-learning skills are enhanced through a series of tutorials, coursework, technical reports, and presentations. Nevertheless, even in this case, general courses addressing risk management and risk analysis are delivered in the first year, while the second year is mostly devoted to specialization in a specific disciplinary area.

Table 5.5: Scheme of the two-year master's program on Reliability, Availability, Maintainability and Safety offered in AY 2018–2019 by NTNU Trondheim (Norway).

Reliability, Availability, Maintainability, and Safety Two-year master's, full-time, in presence
First year
Safety and reliability analysis RAMS engineering and management Maintenance management Risk management in projects Risk analysis Elements of models engineering Experts in a team Select one course from industrial systems engineering; applied statistics; methods and tools in safety practice; subsea production
Second year
RAMS assessment of critical systems Select one course from condition monitoring and maintenance optimisation; safe operation and maintenance; traffic safety and risk evaluation; dependability and performance design RAMS Specialization Project
Master's thesis (30 ECTS) to be selected within a specific specialization area

Table 5.6: Scheme of the two-year two-semester master's program on Safety, Risk, and Reliability Engineering offered in AY 2018–2019 by Heriott-Watt University (UK).

Safety, Risk, and Reliability Engineering **Two-year two-semester master's, distance learning**
Data analysis and simulation
Risk assessment and safety management
Environmental impact assessment
Safety, risk, and reliability
Human factors methods
Systems reliability
Fire, explosions, and process safety
Learning from disasters
Master's thesis (30 ECTS) to be selected within a specific specialization area

Similar educational schemes are delivered by other universities worldwide: for example, Table 5.6 lists the M.Sc. program offered by Heriott-Watt University in Edinburgh [6]. Also in this two-year two-semester online program, fundamental knowledge on data analysis and statistics is delivered together with general elements on risk assessment, including in this case also a specific focus on environmental risk assessment. As usual in the two-semester programs, the focus on a specialization area is obtained by an extended thesis project, usually carried out during an internship in a company.

5.2.3 Process safety management

A number of educational programs offered as second-cycle initiatives (M.Sc. or professional masters), while addressing risk management, focus on specific professional profiles needed for ERManagers in single specialization areas. For example, a number of initiatives address risk management in the chemical industry. Such educational programs, in the area of process safety and loss prevention, are conceptually similar to the above, providing general concepts on risk management in the first period and a specialization on process safety in the second part of the program. For example, the Master of Safety Engineering offered by KU Leuven (Belgium) is a one-year program mostly dedicated to process safety and safety in the process industry [7]. As shown in Table 5.7, the first part of the course addresses introductory subjects as safety engineering, prevention policy, and safety management systems. More technical topics are then introduced, such as safety of products, processes and installations, fire and explosion safety, and so on. The students may then select among two alternative specialization areas, addressing process safety and prevention, respectively. A specific certification required at national level to work as safety inspector is obtained

Table 5.7: Scheme of the one-year master's program on Safety Engineering offered in AY 2018–2019 by KU Leuven (Belgium).

Safety Engineering **One-year master's, full-time, in presence (two-year part-time option available)**
Required courses (23 ECTS)
Explosion safety Qualitative risk analysis techniques Safety of chemical and biological products and of chemical processes Safety aspects of industrial installations Prevention policy and safety management systems
Alternative specialization areas (22 ECTS)
Process safety: provides students with a detailed knowledge of technical and managerial process safety concepts with regard to the whole life cycle of a production plant Prevention: focuses on occupational safety and health-related issues (Certificaat Preventieadviseur Niveau 1) is obtained by completing this option)
Master's thesis (15 ECTS) to be selected within a specialization area

by students completing the "prevention" option. The programme can be completed normally in one year on a full-time basis. However, due to the certificate delivered, the program often attracts working professionals, and a two-year part-time option is also available.

As an example of other similar programs addressing process safety offered worldwide, Table 5.8 reports the scheme of the Safety Engineering Master offered by the MKOC Process Safety Center at Texas A&M University [8]. Table 5.9 reports

Table 5.8: Scheme of the one-year master's program on Safety Engineering offered in AY 2018–2019 by MKOC Process Safety Center at Texas A&M University (USA).

Safety Engineering **One-year master's, full-time, in presence (two-year part-time option available)**
Required courses (16 US credits)
Process safety engineering Industrial safety engineering Quantitative risk analysis Fire protection engineering Statistics
Electives, internships, directed studies, and research (16 US credits)
The remaining 16 US credits are obtained through research (no more than eight US credits), directed studies, internships, and/or elective courses subject to the approval of the program advisor.

Table 5.9: Scheme of the 18-month master's program on Process Safety offered in AY 2018–19 by Universiti Teknologi Petronas (Malaysia).

Process Safety **18-month master's, full-time, in presence (36-month part-time option available)**
Required courses (17 points, approximatively 40 ECTS)
Principles of process safety management Principles of hazard analysis and risk management System safety engineering Safe design and operation Human factors
Required courses (13 points, approximatively 30 ECTS)
Three courses among the following: fire and explosion Engineering; industrial hygiene engineering; quantitative risk assessment; process plant integrity and reliability; incident prevention and emergency management; environmental risk and impact assessment; corrosion engineering Two courses among the following: project management; engineering economy; strategic management
Project (industrial/research) (10 credits – approximatively 20 ECTS)

the outline of the master's in process safety offered by Petronas Technical University in Malaysia [9].

A common educational pattern with the KU Leuven initiative may be recognized: as evident from Tables 5.8 and 5.9, part of the required courses address risk management, human factors, and hazard analysis, as in Leuven (see Table 5.7). A second category of courses address the technical elements required to manage risk in the specialization area addressed, which in the process safety domain mostly concern fire protection, process safety, and safety in design.

5.2.4 Disaster risk management

A specialization area of increasing importance in the present international context is disaster risk management. Although this professional profile is somehow between social sciences and engineering, still in recent years the need of operational managers, thus with a higher knowledge of technical issues, was more and more perceived by public administrations responsible of first response and civil protection services, as well as by companies potentially affected by large scale events.

Table 5.10 lists an example of a M.Sc. program addressing this specific educational profile, offered by University College London (UK) [10]. As evident from the table, the basic elements of risk management are present, but a higher emphasis is

Table 5.10: Scheme of the one-year master's program on Risk, Disaster, and Resilience offered in AY 2018–2019 by Univesity College London (UK).

Risk, Disaster, and Resilience **One-year master's, full-time, in presence (two-year part-time option available)**
Required courses (45 ECTS)
Integrating science into risk and disaster reduction Natural and anthropogenic hazards and vulnerability Emergency and crisis planning Emergency and crisis management Risk and disaster reduction research tools Research proposal and appraisal
Optional modules (15 ECTS)
Two courses among the following options: Conflict, Humanitarianism and Disaster Risk Reduction; Post Disaster Recovery; Adapting Cities to Climate Change; Disaster Risk Reduction in Cities; Earthquake Seismology and Earthquake Hazards; Decision and Risk (Statistics); Risk and Contingency Planning (Security and Crime Science); Risk Power and Uncertainty (Anthropology); Climate Risks to Hydro-Ecological Systems; Perspectives on Terrorism (Security and Crime Science)
Master's thesis (30 ECTS)

devoted to introduce tools and knowledge to manage the social issues related to natural and technological disasters. The program is clearly interdisciplinary and not focused on engineering competences (that may be obtained selecting specific optional courses and a related thesis topic).

Table 5.11 lists an example of a recent M.Sc. program still addressing risk management but more focused on engineering disciplines [11]. The program is offered by University of Padua (Italy) and is similar to others offered in other Italian universities (e.g., Rome "La Sapienza" [12], Genoa [13], etc.). The program is more focused on the technical issues related to safety engineering in the framework of either natural or technological disasters, and therefore devotes less attention to risk management and related organizational, social, and economic factors. Thus, the professional profile addressed is more oriented to first response and prevention rather than to the management of the social and economic issues related to accidents and disasters.

5.2.5 Environmental risk management

In recent years, the concept of ERM has expanded, and the need of professional profiles able to manage risk was also perceived outside the more conventional industrial sectors where such professional profiles were first required. For example,

Table 5.11: Scheme of the two-year master's program on Civil and Industrial Safety Engineering offered in AY 2018–2019 by the Univesity of Padoa (Italy).

Civil and industrial safety engineering **Two-year master's, full-time, in presence**
Required courses (57 ECTS)
Legislation on occupational health and safety Fire prevention Risk analysis Structural safety Construction safety Manufacturing safety
Optional modules (39 ECTS)
Select among two options a) Industrial safety: environmental and safety management; electric safety; structural reliability; safety in energy production and distribution; laboratory of fire safety; safety of transportation and critical infrastructures b) Civil protection: laboratory of fire safety; hydraulic risk; analysis and assessment of geotechnical risk; seismic risk Internship
Internship (9 ECTS)
Master's thesis (15 ECTS)

Table 5.12 lists the outline of an M.Sc. program, offered at University of Tampere, in Finland, addressing the formation of a risk manager profile in the specialization area of environmental risk management in the framework of product design and product lifecycle analysis [14]. The program focuses on two main cross-curricular themes, risk management and circular economy, also illustrating engineering solutions that are, however, presented only as one component in solving the challenges of circular economy. Furthermore, knowledge is provided on technologies for the reuse and recycling of materials, material handling, and the understanding of risks involved with harmful chemicals. Thus, even if the importance given to economic and social disciplines is higher in this program, a number of common features (e.g., risk management, economics, system safety and operations, etc.) may be recognized as fundamental elements in the education provided.

The examples provided show very different educational programs aiming at the formation of an ERManager. Even if the specialization areas are wide, ranging from maintenance to circular economy, some basic elements are always present in the programs, independently on their duration and of the specialization area addressed.

Table 5.12: Scheme of the 18-month master's program on risk management and circular economy offered in AY 2018–2019 by the Univesity of Tampere (Finland).

Risk management and circular economy **18-month master's, full-time, in presence**
Required courses (55 ECTS)
Roadmap for risk management and circular economy Risk management in international operations Megatrends in circular economy Financial management Academic research and practises Recent development in circular economy Leadership for sustainable change Synergy using Piñata method Risk management case Global operations management Communication and conflict
Electives (5 ECTS)
Master's thesis (30 ECTS)

In particular:
- fundamentals of probabilistic calculation and of statistics;
- risk and hazard analysis and assessment;
- management and decision making; and
- economic and social issues related to risk management.

Other concepts frequently delivered in such initiatives concern human and organizational factors and systems engineering.

It is easy to recognize in all the program examples provided the presence of the four categories of skills described in Section 5.1. Moreover, when considering Figure 3.1, one may recognize that most of the elements present in the figure are actually delivered in the different programs discussed above. Cleary enough, depending on the focus, the extension, and the expected attendees, the emphasis given to each of the different elements is different in the different examples provided. Moreover, the main disciplinary area responsible of the educational program (engineering, social sciences, and economic sciences) also influences the balance of the different elements within the educational program. Nevertheless, the examples provided suggest that shared concepts are present in the academic education of ERManagers worldwide, well in accordance with the theoretical elements provided in Chapter 3.

5.3 Illustrative courses

As discussed above, the education of a risk manager can only be carried out by a life-long process; thus, it cannot be limited to a single course.

However, within all programs and educational initiatives, some common key features are present, addressing the specific content of ERM, independent of the specific discipline or specialization area.

It is interesting to find how this common track is always also present in rather different programs and educational initiatives.

An example is provided in Table 5.13, showing the contents of courses related to "risk analysis" in five universities in different countries and/or different disciplinary areas. The courses compared have a different extension (the equivalent number of ECTS, calculated in the table, ranges from 3 to 10). Even if the comparison may be somehow affected by this element, the presence of a number of common elements in the syllabi may be recognized. In particular, risk metrics is introduced, risk calculations are described, and risk acceptance is discussed. Several tools are also common among the different courses (e.g., tools for hazard identification as hazard and operability studies (HAZOP), fault tree analysis for risk calculations, etc.).

Obviously, many other similar examples may be found in academic and professional programs offered worldwide. The presence of these common features in very different contexts worldwide demonstrates that ERM is a mature discipline, with some recognized basic concepts. Thus, even if the safety courses and safety programs are usually clustered and were set-up accounting for local constraints related to the specific local or national context where they are delivered, an international framework and worldwide recognized founding concepts and tools are present. The elements in Figure 3.1 and the specific disciplinary tools developed for their implementation thus are nowadays disciplinary elements shared worldwide within the professional competence required to an ERManager and within the educational programs addressing such professional profile.

5.4 Conclusions

The presence of a common educational framework for an ERManager is clearly evidenced by both the theoretical framework exposed in Chapter 3 and the examples provided in this chapter. However, an ERManager needs an interdisciplinary education that includes both technical skills and managerial and operational capacities. When issuing an educational program, in particular in academia, a primary educational area is usually present, which in general corresponds to that of the university department, school, or college offering the program. This results in differences that somehow bias the importance given to the different skills. Usually, when a program

Table 5.13: Comparison of the content of "risk analysis" courses delivered in five different universities, in different disciplinary contexts, and in different countries. The courses are offered within the programs described in the tables reported above.

University	ECTS	Syllabus
Texas A&M University (USA)	6	Fundamental concepts, techniques, and applications of quantitative risk analysis and risk-informed decision-making for all engineering fields; practical uses of probabilistic methods demonstrated in exercises and case studies from diverse engineering areas (see Table 5.8 for details on the program including this course)
Heriott-Watt University (UK)	7.5	The concept and perceptions of hazards and risk; risk attitudes and impact on decision-making; interpretations of probability; quantitative and qualitative aspects of risk; modeling of decision-making under conditions of risk; Inherent Safety; HAZOP; safety management systems such as BS EN ISO 18000 series and other standards; application of safety management systems to failed systems and as a preventative tool (see Table 5.6 for details on the program including this course)
Stavanger (N)	10	The course covers basic topics in risk analysis and risk management. What is risk and risk analysis? How to plan, execute, and use risk analysis?Risk perspective; uncertainty analysis; risk analysis methods; risk management principles; cost–benefit analysis; cautionary and precautionary principles; risk acceptance criteria; ALARP; decision-making under uncertainty (see Table 5.1 for details on the program including this course)
KU Leuven (B)	3	The different qualitative techniques that are available and can be used for the identification and assessment of risks are described; introduction to hazard identification; collection of data; use of check lists; experience; historical evidence; HAZOP; bow tie analysis; risk matrix approach; dow fire and explosion index (see Table 5.7 for details on the program including this course)
University of Padua (IT)	6	Introduction to the course and relevant legislation. Definition of risk, individual and societal risk; risk acceptability; methodologies for risk assessment; risk identification; accident scenarios and accident chains; HAZOP; what-if; fault tree analysis; FMEA; human factors; models for the calculation of accident consequences; explosive atmospheres (ATEX); risk-based inspection; quantitative area risk assessment (see Table 5.11 for details on the program including this course)
NTNU Trondheim (N)	7.5	Definition and discussion of basic concepts of risk analysis. Risk metrics; risk acceptance criteria; qualitative and quantitative methods for risk analysis, like preliminary hazard analysis, HAZOP, fault tree analysis, and event tree analysis; analysis of human errors and organizational factors; barrier analysis; data sources, and uncertainties; rules, standards, and guidelines; risk reduction and cost/benefit analysis. Use risk analyses within different industries and applications (see Table 5.5 for details on the program including this course)

for the education of an ERManager is offered by a department or a school of engineering, technical skills are the prevailing part of the program; when the program is offered by a department or school of social sciences, human and organizational factors gain importance and visibility in the program; when it is offered by a department or school of economics, the management and economic implications of decision making are usually emphasized. The balance among these three disciplinary areas is thus far from harmonized in educational programs addressing ERM, even when the same specialization area is addressed.

All Universities, when issuing a program, discuss contents with local or national stakeholders (e.g., in the specific case companies in the specialization area addressed, control authorities or public authorities involved in the risk management process, etc.), and usually have no difficulties to include in the program the local requirements deriving e.g. from specific national regulations or technical standards. However, in the case of ERM education, besides local requirements, an international shared educational profile is needed by most companies, since supply chains are usually far more extended than the national dimension, and the attempt of introducing shared practices among production or operational sites in different countries is a pressing requirement of global markets.

Due to the interdisciplinary education required by ERManagers, and to the less consolidated educational profile with respect to monodisciplinary profiles, such as mechanical engineers or economists, it is usually more difficult for a single university to design a program responding to the educational needs at international level.

Therefore, a clear progress would come from the elaboration of an unambiguous set of harmonized core contents in the education of ERManagers among different countries. This process may be based on the activation of a global network of universities, allowing the setup of a harmonized consultation of stakeholders extended at transnational level.

Such process may be a premise to a more pronounced harmonization of educational programs of an ERManagers, adding a common international profile to the specific features of the program, thus better addressing the global competence market.

Accreditation would be perhaps the best tool to support such process. Accreditation bodies for academic programs in the United States, in the United Kingdom, and in other countries actually identify core contents and mimimum number of hours or credits required for specific educational programs [15, 16]. No accreditation body currently addresses ERManager's educational profiles, confirming the difficulty currently experienced in defining an international benchmark for the educational requirements pertaining to this professional profile. The setup of a specific accreditation for ERM academic educational programs would be a relevant step toward the clear identification of the minimum requirements currently needed for ERM education.

When considering the progress in ERM education, besides the present requirements, future needs should also be considered.

Clearly enough the progress of technology and the changes in organizational systems require a continuous update of the specific content of the courses delivered within ERM educational programs. Extended stakeholders consultation is carried out by universities nowadays to update and check the adequacy of their programs. This allows assessing the adequacy of current contents shared by most educational programs addressing ERM.

Nevertheless, besides such a "standard" update, shared by any educational program, more important changes may be considered in the specific context of ERM education. Actually, information technologies are pervading every day more and more in any industrial sector. Global policies launched in specific areas, as the Industry 4.0 initiative carried out by EC, are stimulating such process. The change in the manufacturing scenarios caused by IT may be wide and to date are mostly considered while revising the education of IT specialists and engineers dealing with the specialization areas concerned.

The changes in the production systems introduced by the progress of IT are possibly the more important change that will take place in the near future in industrial systems and in the society. The pervasive role of IT will influence the risk related not only to industrial production, but also to everyday life (e.g., consider unmanned cars). This will have a revolutionary impact on ERM, with implications that in part still need to be fully understood. Although a few programs introduce elements of cybersecurity among ERM education, risk management of cyber–physical interaction is perhaps one of the main areas that is absent in the present education of an ERManager.

More generally, the cyclic and dynamic nature of risk requires the academic and professional education of ERManagers to explicitly address the continuous and lifelong update of technical and managerial skills required to deal with technology-related emerging and ever changing risks.

Strategies based on risks are best solutions to problems having a relevant complexity and some components of uncertainty. If the two most important risk criteria, likelihood of occurrence and severity, are relatively well known and little uncertainty is left, the traditional risk-based approach appears reasonable. Most of the education profiles tend to satisfy this concept.

If uncertainty plays a large role, in particular, indeterminacy or lack of knowledge, this approach reveals some flaws. Judging the relative severity of risks on the basis of uncertain parameters does not make much sense. When uncertainty is the main player, management strategies belonging to the precautionary management style are required. Enhanced education should move forward to this direction as the precautionary approach has been the basis for much of environmental and health protection regulations.

References

[1] University of Stavanger (2018). Risk Management. URL: https://www.uis.no/studies/master-s-programmes-in-english/risk-management/. Last accessed 18.11.2018.

[2] University of Aarlborg (2018). Risk and Safety Management Master. URL: https://www.en.aau.dk/education/master/risk-safety-management. Last accessed 18.11.2018.

[3] Mines ParisTech (2018). Advanced Master MS MRI - Industrial Risk Management. URL: http://www.mines-paristech.eu/Prospective-student/Post-Master-Degree/Advanced-Master/Advanced-master-details/&?catalogid=659. Last accessed 18.11.2018.

[4] ETH (2018). Risk Studies Certificate. URL: http://www.riskcenter.ethz.ch/education/risk-studies-certificate.html. Last accessed 18.11.2018.

[5] NTNU (2018). M.Sc. in Reliability, Availability, Maintainability and Safety. URL: https://www.ntnu.edu/studies/msrams. Last accessed 18.11.2018.

[6] HWU (2018). Safety, Risk and Reliability Engineering, MSc.URL: https://www.hw.ac.uk/study/uk/postgraduate/safety-risk-reliability-engineering.htm. Last accessed 18.11.2018.

[7] KU Leuven (2018). Master of Safety Engineering (Leuven). URL: https://onderwijsaanbod.kuleuven.be/opleidingen/e/CQ_50268884.htm#activetab=diploma_omschrijving. Last accessed 18.11.2018.

[8] TAMU (2018). Safety Engineering. URL: https://engineering.tamu.edu/academics/degrees/seng.html. Last accessed 18.11.2018.

[9] UTP (2018). M.Sc. in Process Safety. URL: https://www.utp.edu.my/Admission/International-Applicants/Postgraduate-Programme/SitePages/MSc%20in%20Process%20Safety.aspx. Last accessed 18.11.2018.

[10] UCL (2018). Risk, Disaster and Resilience MSc. URL: https://www.ucl.ac.uk/prospective-students/graduate/taught-degrees/risk-disaster-resilience-msc. Last accessed 18.11.2018.

[11] University of Padua (2018). Ingegneria della Sicurezza Civile ed Industriale. URL: http://is.dii.unipd.it/. Last accessed 18.11.2018.

[12] UniRoma1 (2018). Safety and Civil Protection Engineering. URL: https://corsidilaurea.uniroma1.it/en/corso/2016/28227. Last accessed 18.11.2018.

[13] UniGE (2018). M.Sc. in Engineering for Natural Risk Management. URL: http://www.cens.unige.it/en/courses-in-the-savona-campus/m-sc-in-engineering-for-natural-risk-management/. Last accessed 18.11.2018.

[14] TAMK (2018). Risk Management and Circular Economy. URL: http://www.tamk.fi/web/tamken/risk-management-and-circular-economy-master. Last accessed 18.11.2018.

[15] Dee, S.J., Cox, B.L., Ogle, R.A. (2015). Process Safety in the Classroom: The Current State of Chemical Engineering Programs at US Universities. *Process Safety Progress*, 34, 316–319.

[16] IChemE. (2015). Accreditation of Chemical Engineering Programmes – A Guide for Higher Education Providers and Assessors. IChemE.

6 Conclusions and perspectives

6.1 Perspectives on risk management education

Our society is more and more characterized by uncertainty: hazards and their associated threats, and a proliferation of different risks. We so frequently defend ourselves against large and small potential harms that we practise risk avoidance almost without thinking: washing hands, buckling seatbelts, wearing helmets are but a few risk-related behaviors that we practice without much conscious reflection. Sometimes the risks that we take are voluntary, while others are imposed upon us. Risks can be personal or public, they can be local or global like climate change and pollution, and are therefore increasingly shared. The new risk trends are emerging simultaneously and will compel the forward-looking organizations to adopt or adapt new or existing risk management systems and processes. In this framework, ERM functions in companies need to change their nature. While engineering and technical risk management was established in several industrial organizations with the primary aim to focus on regulatory compliance, the fast-changing and ever-evolving nature of risk now requires that risk management is embedded in any activity, making safety second nature [1]. Today, risk management is driven by the corporate risk function. In the near future, we could expect that embedded processes and new automation will have engaged the business in managing risk more fully. The purpose of the risk management framework is to assist the organization in integrating risk management into significant activities and functions. The effectiveness of risk management will depend on its integration into the governance of the organization, including decision-making. Risk management should be a part of, and not be separate from, the organizational purpose, governance, leadership and commitment, strategy, objectives, and operations.

One of the key issues in modern risk management is education. The value of risk management lies not much in the ways it serves as means to investigate social responses to particular forms of possible danger, but more in the ways in which it might sensitize us to use the language in current policy debates surrounding individual responsibilities. This should invite us to have a prospective attitude to reimagine ourselves as subjects with a fluid set of aspirations within the context of risky futures.

Today's students, consumers, and businesses are accustomed to personalization through social media and to rapid fulfillment through e-commerce. They will expect the same kind of near-instantaneous service and customized products from education. A special attention should be paid to the positive aspects of mindfulness and creativity that make teaching innovative, exciting, and challenging.

A major transformation happening is digitalization. Digital risk transformations are already a reality at different business sectors but very scarcely taught in any education profile. It will be imperative for the risk function to accelerate its

https://doi.org/10.1515/9783110344578-006

digitization efforts and its education, since it will be increasingly hard to stay analog while customer-facing activities and operations race ahead into being completely digital. A digital transformation for risk would mean a number of changes. Risk would capture and manage information from a broader and richer set of data, looking into nontraditional sources like business-review ratings online. It would automate the processes it controls, and work with others to do the same for decision-heavy processes. It would use advanced analytics to further improve the accuracy and consistency of its models, in part by greatly reducing the biases. Risk would review and reshape its mandate and role to capitalize on its ability to provide faster, more forward-looking, and deeper insights and advice. It would alter its organizational setup, as well as its culture, talent, and ways of working.

The education of engineering risk managers should address the future concerns of modern management of risk. The latter will need to be able to identify and address new risks quickly, be more agile and modular, deliver new technology and techniques rapidly, and work increasingly in partnership with finance, operations, and the businesses. ERManagers need to become more familiar and aware of the behavioral aspects of risk management. In addition, tools and techniques may need to evolve to reflect the changing nature of the risk framework and landscape. These changes will require them to recruit, develop, and retain staff with skills that differ significantly from those that are found in organization today. These skills should at least be the concern of modern education at all levels: undergraduate, graduate, and continuing education.

6.2 New technologies and new delivery formats for education and networking

The revolution in the role of ERManagers required by the dynamic framework of risk evolution, and the constantly renewed competences that need to be provided in ERM education may be supported by the simultaneous revolution that is investing in technologies and models for education.

Digitalization affected since a number of years the education sector. As a result, a variety of ITtools is now available to support new delivery formats that provide competencies within the educational process. Such innovative technologies are likely to have their more important impact on the professional and continuing education models, since a number of effective delivery formats supported by new technologies are now available for distance learning and time-flexible blended courses, which are the more suitable formats for continuing education of professionals. Table 6.1 lists some key features of e-learning formats that are of particular importance to the education of ERManagers in particular when professional education is considered. In fact, e-learning is able to widen the target audience, allowing learners who may not attend conventional classroom training to participate in educational programs taking

Table 6.1: Some key characteristics of e-learning formats of particular importance for the delivery of educational programs aimed at the formation of ERManagers.

Feature	Description
Distance learning	Allowing the delivery of an educational program to geographically dispersed learners with limited time and/or resources to travel
Self-paced study	Allowing the personalization of learning paths based on learners' needs and using case studies, simulations, and other IT-based delivery formats
Part-time learning	Allowing the participation of learners busy with work or family commitments that do not allow them to attend courses on specific dates with a fixed schedule
Linguistic differences	Facilitating the management of linguistic differences, mitigating the difficulties present in real-time communication typical of conventional educational models when delivered in a language different from the mother tongue of the learners
Quality in education	E-learning delivery formats may mitigate the differences among the specific instructors, thus promoting a standard quality in the delivery of the program
Cultural diversity	Distance learning and self-paced instructions may ease the management of cultural diversity, for example, promoting the involvement of learners limited from participating in classroom sessions because of cultural or religious beliefs

advantage of distance learning or blended learning delivery models. The possibility to introduce asynchronous delivery methods in e-learning formats, allowing self-paced learning, is the key factor contributing to widen substantially the potential participation in educational programs by professional or part-time learners.

Although these advantages apply to any educational field, they assume particular importance in the education of ERManagers due to the importance of professional education in the more frequent educational models for this specific educatonal profile (see Figure 5.1).

It should also be mentioned that the cost of e-learning formats and/or distance learning is substantially lower; thus, companies, in particular, small- and medium-sized enterprises, may be more keen to afford the promotion of such educational programs among their personnel.

Specific tools and technologies are needed in order to produce and deliver e-learning contents. It should be remarked that the advent of e-learning caused the development and dissemination of a number of new and innovative tools that go beyond those traditionally used in conventional education programs.

While slide-editing and word-processing software were the key of conventional delivery formats, and can still be sufficient to create simple learning resources like a presentation or a tutorial, new tools and IT resources specific of e-learning also have the potential to deeply innovate conventional teaching.

Several different tools can be used to produce e-learning contents, depending on which is the nature of the desired final product. Courseware authoring tools may be used to create interactive e-learning contents. These tools allow adding text, graphics, short videos, and material in other formats, also allowing the organization of the material in a framework facilitating the navigation across pages and single lessons. Either stand-alone software packages, incorporating proprietary tools, or tools capable to integration with specific software are now available.

Moreover, universities and other specialized institutions providing educational services are implementing commercial or customized learning platforms to support e-learning. A learning platform is a set of interactive online services that support the delivery of a course. They can be used either for distance learning or to support the delivery of conventional and/or blended courses. Learning platforms provide the learners with an easier access to information, while making available to the learner a number of tools and resources supporting the learning process.

A variety of learning platforms are available nowadays, with different levels of complexity. The more important features of such tools include:

- the possibility to include interactive materials promoting learner engagement;
- an effective management of learning contents, by an easier management of file storage and of access to resources;
- the possibility to manage the delivery and assessment of homework and of individual work delivered by the learners, as well as to include materials for the self-assessment of the learners; and
- the availability and easy access to specific tools and services: forums, blogs, wiki tools, group discussions, and messaging.

Again, even if such features may be beneficial to any educational field, when dealing with the final steps in the education of an ERManager (see Figure 5.1), a huge potential is present for the valorization of the additional potentialities of IT platforms supporting the learning process. Actually, as discussed in Chapter 4, when dealing with the professional educational tier (see Figure 4.2), the use of interactive teaching methods, of case studies, and of other innovative delivery methods is proving of particular effectiveness. Moreover, the development of web contents is, in perspective, a key to overcome the jeopardized competences present in different universities and in different countries, supporting and facilitating the harmonization process of educational programs discussed in Section 5.4.

The widespread of e-learning formats is promoting a revolution in teaching methods. Besides the support provided by IT tools and the advantages that e-learning may have with respect to traditional learning in particular for higher education, the shift from traditional learning to e-learning and/or to blended learning is promoting a change in delivery formats that will reverberate also on conventional, in-presence, full-time educational programs. This is expected,

in particular, when considering blended learning. Actually, two models were identified by Bersin [2] for blended learning:
- Program flow model: Learning activities are organized in a linear, sequential order and learners have deadlines to accomplish the various assignments (the model is similar to traditional training, but with some of the activities conducted online).
- Core-and-spoke model: A conventional course (in person or in remote) is delivered, and a set of supplemental materials and activities are made available to reinforce the main course.

It is evident that both modalities may be implemented and supported by e-learning technologies and may find in IT learning platforms a symbiotic tool to support interaction and a shift towards learner-centered teaching.

The new IT tools support innovative delivery formats, making the adoption of interactive teaching methods easier and at-hand. Interactive teaching and learner-centered teaching are assumed to be of key importance in modern educational programs. Consider, for instance, the main teaching methods:
- *Expositive methods*: Aimed at delivering new information and/or knowledge, they are usually based on lectures, presentations, case studies, worked examples, and demonstrations.
- *Application methods*: Aimed at providing new skills to the learners, they are based on interactive and/or active processes, requiring the learners to complete specific tasks. They are based usually on case studies, scenario-based exercises, role play, simulations, guided research, and project work.
- *Collaborative methods*: Focus on developing the social dimension of learning, thus engaging the participants to share knowledge and perform collaborative tasks. They are usually based on guided discussions and collaborative (group) work.

Table 6.2 shows how the innovative IT tools have the potential to change and/or to be strongly integrated with the traditional delivery formats of such teaching methods. It is clear from the table that the new IT tools significantly widen the tools available for teaching, in particular, when considering applicative teaching methods, being a key element of professional education of ERManagers, as discussed in Chapter 3.
Professional education is heavily based on case studies, scenario-based learning, role playing, and other similar contents typical of applicative teaching methods. The new technologies available nowadays will contribute to support the development of IT-based teaching material, thus considerably extending the potentialities of effective professional teaching in safety and risk management.

The progress with learning platforms will provide progressively more effective distance learning opportunities for professional formation, and will decrease the cost of such instruction, thus increasing its penetration among professionals. Thus,

Table 6.2: Innovative delivery formats allowed by IT tools supported by learning platforms.

Method	Conventional	IT-based supported on learning platforms
Expositive	– "Chalk and Talk" – Presentation supported by slides with no interactivity	– Interactive e-lesson contents based on text, images, audios, animations, and practice – Presentations made by an expert or instructor that are broadcast in real time or recorded for learners to watch at any time – Webinar and virtual classroom: The instructor presents the content to a group of learners who are connected to the platform at the same time. Learners can interact with the instructor, ask questions, and receive feedback using video conference, audio conference, or chat.
Applicative	– Guided examples with the instructor showing the solution to a challenging situation or to a case study – Individual or group assignments or homework. Feedback provided after assessment of delivered paperwork – In person role-play in the presence of an instructor or of a tutor	– Interactive e-lesson contents based on combining animations and operational simulations. Learners may interact and receive feedback from the system. – Virtual classroom supported by application sharing tools. The software allows learners to interact and to practice the case provided. – Experiential simulations based on "branching scenarios." A feedback on learners choices is provided through a follow-up situation in which several further options are possible. – Tutorials where a challenging situation is presented together with the information and tools required to develop an individual solution, which is the required output. Feedback is provided by the tutor during or only at the end of the work. – Role-play using communication tools such as chats, audio or video conferences, and discussion forums.
Collaborative	– In-presence interaction in the classroom during guided discussion – Group work on assignments or homework	– Discussion forums, e-mails, chats, and audio or video conferences for communicating between learner and instructor or tutor, also allowing asynchronous group work. – Wikis and blogs allowing shared work on documents for the elaboration of results.

in perspective, new technologies are expected to play a positive role in increasing the education of ERManagers, and in particular in providing a substantial increase in the opportunities of professional and continuing education, thus supporting the creation of the dynamic, life-long learning environment needed in the framework of ERM, as discussed in Section 5.4.

The availability of learning platforms and IT tools will also likely play a role on the easier preparation and sharing of educational materials and contents [3]. In particular, the IT tools available may support the preparation of experience-based material and the circulation of teaching material among the existing international networks among academia and industry in the field of safety and ERM. The possibility to obtain and use teaching materials supported by learning platforms potentially enhances the role of national and international organizations and networks having the mission to prepare and disseminate teaching materials on safety and risk management.

In perspective, new technologies and new delivery formats may contribute to the enrichment and spreading of an ERM culture, and to the capillary dissemination of suitable teaching material to update and further qualify the available education programs in the framework of ERM, both within academia and in professional education.

6.3 Competences and resources for education

Safety education and training (including for instance security, unsafety, and health issues) and learning about risks in general, and how to manage them, is a lifelong process for every human being. This is just a fact of life. Dealing with risks is dealing with uncertainties and it helps in life and in working conditions to be knowledgeable about the risk decision-making process to achieve the best possible outcomes. Being successful in life is therefore largely dependent on the competences one gains about risk-related decision-making. As in other domains, some people learn much faster about risks, unsafety, and safety than others, and some will never learn. But for everyone, ERM education, training, and learning is very important. The level of the learning, and the pace with which one learns ERM competences, can literally mean the difference between life and death. The past, but certainly the future of safety science and ERM, will therefore be paved with knowledge and know-how.

Risk and safety competences are very diverse, as was demonstrated in the previous chapters. One not only needs to learn from very specific and concrete examples such as near misses, incidents, and major accidents, but also from safety models, theories, and risk knowledge in general. It is essential that people learn to think proactively and long term, besides the much easier and human-nature related reactive and short-term thinking. A course on ERM, starting from all programs in regular education (primary school, high school, and university college or university), and further transforming into professional education in all organizational contexts, is what society truly needs to further advance. If humans get familiar with safety and the engineering risk concept from very early ages, they can learn much more about these topics in higher education. Moreover, it can be expected that much more thorough safety knowledge of all people through regular education will be used in daily life and business to make better decisions and to reduce losses, both on private- and public working levels. Professional education would be much more efficient and could be more focused on the organizational aspects of ERM.

Society has evolved over the past decades toward being ever more involved and interested in safety (avoiding losses), security (terrorist threats), risk (major accidents) , environment (climate change), and health (ageing societies). Society's tools for tackling these problems and moving toward a better future worldwide are education, collaboration, and transparency, in that sequential order. Modern private and public organizations are growing in scale and complexity. The interconnectedness and interdependence of citizens and organizations has never in history been higher. Systemic risks, on top of "regular" smaller scale risks, therefore emerge and threaten organizations and people's welfare as never before. Education, training, and learning about ERM is indeed the only correct answer to deal with this changing societal context. Collaboration and transparency are not essential, but they are very important tools to improve the learning aspect. Humans must be properly educated and trained: they must be made aware of how incidents and accidents occur, what can go wrong, how equipment or systems fail, how human error leads to incidents involving loss of life and property, what models and theories exist, how to do things right, how to make the decisions that lead to long-term profitability, and so on.

The previous chapters have provided a basic understanding of the two-leveled approach (regular education and professional training) and the required knowledge and know-how, skills, and competences needed for understanding and applying ERM and making sound safety-related decisions.

A subject deserving much more attention in education of ERManagers is security, both physical and cyber. The difference between safety and security concerns intentionality: safety deals with unintentionally caused losses, while security tackles deliberately caused losses. A lot of theories, models, and applications that have been developed in the past century from a safety perspective can readily be used for security purposes, or can easily be adapted to the needs of security. Currently, courses on safety usually do not mention or treat security at all, although similar lines of thinking between safety and security exist.

Another important aspect that needs more attention, both in courses offered in regular education and in professional learning, concerns the hypothetical benefits (or avoided losses) gained through safety investments. Indeed, an important way to further improve safety is to understand and use economic analyses for safety investment decision-making. As Reniers and Van Erp [4] indicate, economic analyses, if carried out correctly, almost always show that safety investments are a no-brainer, that is, they should be carried out, and that investing in prevention and avoiding accident costs actually is a business strategy leading to long-term profitability and to sustainable and inherently healthy organizations. Similar to the fact that good decisions should be taken in the fields of financial uncertainties and profit-related operational uncertainties (e.g., production investments and innovation), good decisions should also be taken to avoid operational losses. For the long-term profitability of any organization, production and innovativeness, on the one hand, and safety and security (i.e., avoiding all types of losses), on the other hand, are equally important. Actually, leadership is all about ERM, and it is just idle talk if not the first thing that leaders should learn and understand is that there are two requirements to raise profits: increasing production and innovativeness, leading to direct and tangible profits, and decreasing the numbers and the level of losses, leading to indirect and intangible (but real) profits (so-called hypothetical benefits).

In the same line of thinking, resources for safety education in organizations are often regarded as a cost. As Hollnagel [5] explains, spending significant amounts of time to learn, think, and communicate about safety is usually considered as a cost. Money spent on safety (and safety education) is considered a cost or at least a lost opportunity for investment in other sectors, as productivity. Since it is impossible to prove that safety precautions and investments really worked, and the concept of “hypothetical benefits” is extremely difficult to explain to decision-makers in the sense that they will take them into account in their decisions, and since it is not possible to predict when an accident is likely to happen, calculations are currently biased in favor of reducing safety investments. Moreover, this is something that typically takes place in hard economic times. As Hollnagel [5] also remarks, it defies logic because the probability of an accident is not reduced when economic conditions get worse. Actually, the opposite is true: the accident probability increases during hard times, since people are pressed harder, are less likely to report incidents, are more likely to take chances and changes, and so on. Therefore, this observed thinking phenomenon and accompanying behavioral process is counterintuitive, and can only be countered with the right mind-set. Hence, the importance of education and training for creating the right mind-set, perception, and attitude, and not “merely” for gaining safety knowledge and know-how.

The learning process and the subsequent safety competences should thus encompass several aspects:

i. Theoretical ERM knowledge
ii. The right ERM mind-set, perception, and attitude
iii. Practical ERM know-how
iv. Behavioral ERM skills

Items (i) and (ii) represent long-term oriented person-based safety and can best be taught in regular educational programs. They form the basis for items (iii) and (iv), representing short-term oriented behavior-based safety, being achieved via professional training. All items are absolutely necessary for a successful ERM.

As Nicholas [6] explains, two types of learning exist: technical and adaptive learning. A large part of the challenge for a student seeking to develop any capability (for instance concerning risk and safety) that is based on mind-set is that one is used to thinking about the personal growth in terms of skills, especially "hard" skills that are teachable abilities that are relatively easy to measure and quantify (such as data analysis, computer programming, carrying out risk assessments such as HazOps, knowing and applying the legislation, etc). "Soft" skills, conversely, as communication (speaking, writing, or listening – extremely important for an ERManager), negotiating, teamwork, or critical thinking (rationalizing) are highly subjective and therefore difficult to quantify, and much less easy to learn using the familiar approach whereby somebody who knows teaches someone who does not know. Unfortunately, trying to teach mind-set–based capabilities via the old model of instruction will have no greater impact than does explaining to somebody that if he/she wants to lose weight, he/she needs to eat less food. A lack of understanding is not the problem! Hence, the traditional approach to personal development largely fails when the mind-set change is required.

In summary, there are two different forms of personal change that require a different type of learning, depending on which part of the brain is involved: "technical" changes and "adaptive" changes. The former changes are enabled by the development of skill sets and have a cognitive basis, while the latter changes require a need to transform at a deeper level than purely to develop new skills. Technical changes can be realized using the, approach for effective learning. Adaptive changes must be addressed from the "inside-out", through a process of realization and self-discovery rather than instruction, and impacts the way people perceive the world, the level of self-awareness that people have, the way in which people see the world.

This distinction between technical and adaptive changes obviously has an impact on the learning process. Both changes need to be taught and upgraded throughout a lifetime. But it is essential that the foundation for the right mind-set regarding ERM is laid from an early age onward, and that besides technical changes (risk and safety knowledge) also adaptive changes (perception and attitude regarding risk and safety) are realized via regular education. Professional education should then have the duty to specialize and upgrade both technical and adaptive ERM changes, since flexibility and adaptation are needed in the field of risk and safety.

Regarding the concrete resources that can be employed for teaching in regular and professional courses, a variety of approaches and materials are available (non-exhaustive list):

- Literature of all kind (books (a.o., textbooks, specialized books), academic and professional articles, grey literature, legislation, guidelines, codes of good practice, material safety data sheets, etc.)
- Audiovisual aids (pictures, schemes, schedules, maps, charts, etc.)
- Movies, recordings, and the alike

- Other communication aids, such as social media (Twitter, Facebook, LinkedIn, etc.)
- Case studies and accident investigation reports
- Internet websites (such as from the Chemical Safety Board for instance)
- Guest lectures with specific knowledge, messages, and/or stories
- Actors that can do role-playing
- Field visits
- Discussion panel between experts
- Debate between students about a certain subject, moderated by the teacher
- Showcases (either in a room or in the field), laboratory examples/sessions

As the final thought of this book, one should realize that in every large group of students, whatever their age, there will always be very quick learners and very slow learners. Most of them will usually be average learners. The most important thing a teacher should remember can be summarized in a wise English proverb saying "You can lead a horse to water but you can't make him drink," meaning in this case that you can give someone an opportunity to learn about risk and safety to become an excellent ERManager, but you cannot force him/her to take it. Motivation to become an excellent ERManager should also come from the personal understanding of the importance of such role in society.

References

[1] Kletz, T.A. (1998). Making safety second nature. *Process Safety Progress*, 17 (3), 196–199.

[2] Bersin, J. (2004). *The Blended Learning Book*. San Francisco (USA): Pfeiffer.

[3] Nash, S., Moore, M. (2014). *Moodle Course Design Best Practices*. Birmingham (UK): Packt Publishing Limited.

[4] Reniers, G.L.L., Van Erp, H.N. (2016). *Operational Safety Economics. A Practical Approach Focused on the Chemica Land Process Industries*. Chichester (UK): John Wiley and sons.

[5] Hollnagel, E. (2014). *Safety-I and Safety-II. The Past and Future of Safety Management*. Farnham (UK): Ashgate.

[6] Nicholas, M. (2017). *The Little Black Book of Decision-making*. Chichester (UK): Capstone (Wiley).

Acronyms

ABET	Accreditation Board For Engineering And Technology
AIChE	American Institute Of Chemical Engineers
ALARA	As Low As Reasonably Achievable
ALARP	As Low As Reasonably Practicable
AY BSc	Academic Year Bachelor Of Science
CEAB	Canadian Engineering Accreditation Board
Ceng	Chartered Engineer
CFSE	Certified Functional Safety Expert
CPD	Continuous Professional Development
CPI	Chemical Process Industry
Dechema	Deutsche Gesellschaft Für Chemisches Apparatewesen, German Society For Chemical Apparatus
ECCE	European Congress Of Chemical Engineering
ECTS	European Credit Transfer Scale
EFCE	European Federation Of Chemical Engineering
ERM	Engineering Risk Management
ERManager	Engineering Risk Manager
ETH	Eidgenössische Technische Hochschule, Swiss Federal Institute Of Technology
HAZOP	Hazard And Operability Analysis
HSE	Health & Safety Executive
IChemE	Institution Of Chemical Engineers
INSAG	International Nuclear Safety Advisory Group
ISO	International Organization For Standardization
IT	Information Technology
KPABC	Knowledge Perception Attitude Behavior Consequences
KU	Katholieke Universiteit, Catholic University
MOOC	Massive Open Online Course
MSc	Master Of Science
NAIIC	Nuclear Accident Independent Investigation Commission
NORSOK	Norsk Sokkels, Norwegian Shelf's Competitive Position
NPP	Nuclear Power Plant
NTNU	Norwegian University Of Science And Technology
OECD	Organisation For Economic Co-Operation And Development
OHSAS	Occupational Health And Safety Assessment Series
OJT	On-The-Job Training
PEng	Professional Engineer
PhD	Philosophiæ Doctor, Doctor Of Philosophy
PPSE	Professional Process Safety Engineer
RM	Risk Management
SAChE	Safety And Chemical Engineering Education
TEPCO	Tokyo Electric Power Company
TMI	Three Mile Island
USNRC	Us Nuclear Regulatory Commission

https://doi.org/10.1515/9783110344578-007

Index

ABET 55, 59
Academic programs 70
Acceptance 11
Accidents 81
Accreditation 55, 62, 63, 86
Adaptive learning 10
Adverse event 21
AIChE 59
ALARA 29
ALARP 29
Application methods 5
Appraisals' 47
Asset integrity 61
Assimilation 61
Assistantship 48
Asynchronous delivery methods 3
Attitude 21, 51
Auditor 6
Availability 77
Awareness 9, 51

Behavior-based safety 19, 26, 57
Bias 84
Blended learning 3, 4, 46
Blogs 4
Bologna educational system 55
Bologna process 73

Capability 55
Case studies 5, 44
Case-histories 70
Catastrophe 6
CEAB 55
Chemical engineering programs 56
Chemical industry 78
Chernobyl 7
CHESS 15
Circular economy 82
Classic risk management 35
Classroom 42
Coaching 48
Collaboration 59
Collaborative methods 5
Commitment 1
Communication 10, 49
Competence 2, 4, 7, 24, 25, 72
Competence providers 69
Competencies 60
Competency 55, 56
Compliance 3
Conference 43
Conformance 47
Continuing education 2, 6, 38
Continuous learning 54
Continuous Professional Development 53, 54
Conventional teaching methods 70
Core competences 55
Core contents 86
Cost-efficient 5
Courseware authoring tools 3
Credits 86
Crisis management 75
Cultural diversity 3
Curricula 56, 62
Curriculum 11, 56
Customized products 1
Cyber-physical interaction 87
Cybersecurity 87
Cyclic 87

Data analysis 78
Dechema 56
Decision-making 1, 7, 8, 83
Deficiencies 6
Delivery formats 2
Demonstration 45, 48
Digital risk 1
Digitalization 18
Disaster 10, 81
Disaster risk management 73
Discipline 84
Discussion 43
Distance learning 2, 3, 4, 5, 73
Dynamic 4, 87

Economic analysis 8
Economic sciences 83
Economic skills 30, 69
ECTS 84
Education 1, 2, 9, 18, 20, 25, 26, 33, 52, 76
Education 6

https://doi.org/10.1515/9783110344578-008

Education model 53
Education profile 2, 29, 30
Educational approaches 18
Educational area 84
Educational backgrounds 30
Educational process 2, 69
Educational programs 3, 4, 5, 9
Educational route 31
Educational schemes 78
Educational tool 23
EFCE 59
Effectiveness 1, 4
e-learning 2, 3, 4, 5, 46
Engineering competences 81
Engineering disciplines 76
Engineering risk management 1, 2, 6, 7, 8, 9, 31, 70
Engineering Risk Manager 2, 4, 8, 10, 11, 69
Engineering solutions 82
Environmental risk assessment 78
Environmental risk management 73, 81
Evolution of safety 15
Experience-based learning 45
Experimental safety 19
Expositive methods 5
Exposures 1

Failures 2
Field visits 46
Films 45
Fire and explosion safety 78
Fire protection 80
First response 81
Five senses 41
Flexibility 21
Flipped classroom 45
Fukushima 8

Global network 86
Governance 1, 3
Group visits 46
Guidelines 22

Hard skills 10
Harmful chemicals 82
Harmonization 4
Harmonized 86
Hazard identification 84
High school 17
Holistic 6
HSE 22
Human and organizational factors 83
Human behaviour 3
Human factors 80
Human role 7
Hypothetical benefits 8, 9

IChemE 55
In-basket 45
Incidents 7, 8, 9
Industrial risk management 73
Industrial systems 87
Industry 4.0 87
Industry-based research program 54
Inherently safer design 56
Inspection 6, 54
Instructions 6
Interactive teaching 5
Interactive teaching methods 4, 5, 70
Interdisciplinary 64
Internship 48, 54, 70
Intolerance 6
Investment 5
ISO 19011 24
ISO 22
ISO 31000 3, 22, 24
ISO 45001 22, 24, 25
IT technologies 87

Job rotation 48
Job-instruction 48

Kaizen 5
Key features 84
Knapsack 5
Know-how 35
Knowledge 9, 21, 35, 51, 55, 77
Knowledge owners 70
KPABC 20, 25

Laboratory training 45
Language 49
Leadership 1, 3, 9, 21
Learner-centered teaching 5
Learners 2, 4, 5, 11, 72
Learning 3, 4, 5, 7, 8, 9, 10

Learning aids 63
Learning cases 70
Learning characteristics 51
Learning contents 4
Learning from inspections 53
Learning methods 60
Learning platforms 4, 5, 7
Learning process 69, 70
Learning pyramid 69
Learning pyramid model 57
Lecturing 43
Lessons learnt 70
Life-long learning 6
Life-long process 84
Likelihood 4
Linguistic differences 3
Local requirements 86
Long-distance learning 46
Long-term profitability 9
Loss prevention 78
Losses 8, 9

Maintainability 77
Major accidents 7, 8
Management 2, 4, 17, 21, 24, 86
Managerial skills 52
Manufacturing 87
Massive Open Online Course 46
Material handling 82
Mathematics 74
Mentoring 39
Mindfulness 1
Mind-set 9, 10
Monitor 4
Monodisciplinary profiles 86
MOOC' 46
M.Sc. Programs 72

National regulations 86
Natural 81
Natural hazard 10
Near-misses 7
NORSOK 22

Objectives of the education 52
Observation 47
Occupational 24
OHSAS 22
Onagawa 8
On-the-job 47
On-The-Job training 54
Operational losses 9
Operational risk management 75
Organisational 52
Outcomes 51
Outreach 40

Panel discussion 43
Paradigm 64
Part-time 72
Part-time learning 3
Perception 21
Performance 21
Person-based safety 19
Policy 10
Practice 12, 47, 62
Practice by doing 48
Precautionary management 87
Presentation 3, 5
Prevention 81
Primary school 17
Probability 4, 74
Process safety 78
Process safety bilateral collaboration 60
Process safety collaboration 58
Process safety competence 61
Process safety curriculum 59
Process safety education 53
Process safety management 61, 73, 78
Process safety Research 57
Process safety topics 61
Process safety training 61
Product design 82
Product lifecycle analysis 82
Production systems 87
Professional 26
Professional course 36
Professional education 2, 3, 5, 7, 10, 35, 52, 54
Professional experience 72
Professional learning experience 70
Professional Master Programs 72
Professional programs 70
Professional training 54, 61
Professionals 40
Program 11, 53, 60, 62

RAMS 77
Recording 4

Regular education 33, 52
Regulations 11, 63
Regulatory compliance 1
Reliability 73, 75
Reporting 4
Resilience 21
Resources 3, 4, 5, 9, 10
Responsibilities 1
Retention Pyramid 38
Retention rates 41
Review 4
Risk 1, 17
Risk acceptance 84
Risk analysis 33, 75
Risk appetite 2
Risk assessment 29, 32, 78
Risk avoidance 1
Risk awareness 13
Risk calculations 84
Risk function 1, 2
Risk knowledge 7
Risk management 22, 33, 69, 75
Risk management systems 1
Risk management tools 74
Risk manager 29, 37
Risk manager profile 32
Risk reduction 9
Risk trends 1
Risks 1, 2, 7
Role playing 70
Role-playing 44
Root cause 12

SACHE 59
Safety 1, 2, 5, 6, 7, 8, 9, 10, 11, 21, 22, 77
Safety audits 47
Safety barriers 69
Safety courses 52
Safety culture 6, 7, 11
Safety education 7, 9, 52
Safety engineering 78
Safety in design 80
Safety investments 8, 9
Safety knowledge 11
Safety management 11, 73
Safety models 7
Safety of products 78
Safety performance 59
Safety progress 16
Safety revolutions 15
Safety science 7
Safety-I 21
Safety-II 21
Scenario-based learning 5
Second cycle degrees 73
Sectorial needs 75
Security 8, 9
Self-awareness 10
Self-paced study 3
Seveso Directive 54
Shared concepts 83
Shared educational profile 86
Shared practices 86
Showing 45
Simulation 45
Skills 51, 55, 83
Slide-editing 3
Small- and medium-sized enterprises 3
Social issues 81
Social media 1, 11
Social sciences 83
Social skills 30
Soft skills 10
Specialization area 77
Specific skills 30
Stakeholders 3, 86
Standards 23
Statistics 74, 78
Strategy 1, 3
Students 1, 11
Supply chains 86
Systemic risks 8
Systems engineering 83

Teaching 17, 33, 51
Teaching approaches 62
Teaching material 70
Teaching materials 7
Teaching methodologies 56
Teaching others 48
Technical issues 81
Technical learning 10
Technical risk management 1
Technical skills 30, 84
Technical standards 23, 86
Technological disasters 81
Technologies 2, 3, 5, 7